# Intravascular Ultrasound

## From Acquisition to Advanced Quantitative Analysis

EDITED BY

SIMONE BALOCCO

ELSEVIER

Elsevier
Radarweg 29, PO Box 211, 1000 AE Amsterdam, Netherlands
The Boulevard, Langford Lane, Kidlington, Oxford OX5 1GB, United Kingdom
50 Hampshire Street, 5th Floor, Cambridge, MA 02139, United States

**Notices**

Knowledge and best practice in this field are constantly changing. As new research and experience broaden our understanding, changes in research methods, professional practices, or medical treatment may become necessary.

Practitioners and researchers must always rely on their own experience and knowledge in evaluating and using any information, methods, compounds, or experiments described herein. In using such information or methods they should be mindful of their own safety and the safety of others, including parties for whom they have a professional responsibility.

To the fullest extent of the law, neither the Publisher nor the authors, contributors, or editors, assume any liability for any injury and/or damage to persons or property as a matter of products liability, negligence or otherwise, or from any use or operation of any methods, products, instructions, or ideas contained in the material herein.

**Library of Congress Cataloging-in-Publication Data**
A catalog record for this book is available from the Library of Congress

**British Library Cataloguing-in-Publication Data**
A catalogue record for this book is available from the British Library

ISBN: 978-0-12-818833-0

For information on all Elsevier publications
visit our website at https://www.elsevier.com/books-and-journals

*Publisher:* Mara Conner
*Acquisitions Editor:* Tim Pitts
*Editorial Project Manager:* Fernanda A. Oliveira
*Production Project Manager:* Kiruthika Govindaraju
*Cover Designer:* Alan Studholme

Typeset by SPi Global, India

# Intravascular Ultrasound

## From Acquisition to Advanced Quantitative Analysis

# Contents

# Contributors

*The editor would like to acknowledge and offer grateful thanks for the input of all contributors, without whom this first edition would not have been possible.*

**Simone Balocco**
Department of Mathematics and Informatics, University of Barcelona, Barcelona; Computer Vision Center, Bellaterra, Spain

**R. Pawel Banys**
Department of Radiology, John Paul II Hospital; Department of Physics and Applied Informatics, AGH University of Science and Technology, Krakow, Poland

**Stéphane Carlier**
UMONS & CHU Ambroise Paré, Mons, Belgium

**Xavier Carrillo**
University Hospital Germans Trias i Pujol, Badalona, Spain

**Maria Elena de Ceglia**
InspireMD, Tel-Aviv, Israel

**Zhi Chen**
Electrical and Computer Engineering; Iowa Institute for Biomedical Imaging, University of Iowa, Iowa City, IA, United States

**Francesco Ciompi**
Diagnostic Image Analysis Group, Pathology Department, Radboud University Medical Center, Nijmegen, The Netherlands

**Wladyslaw Dabrowski**
Jagiellonian University, Department of Cardiac & Vascular Diseases; Department of Interventional Cardiology, John Paul II Hospital, Krakow, Poland

**Stamatia Giannarou**
Hamlyn Center for Robotic Surgery, Imperial College London, London, United Kingdom

**Joan Antoni Gomez-Hospital**
Interventional Cardiology Unit, Hospital de Bellvitge, Barcelona, Spain

**Josep Lluís Gómez-Huertas**
InspireMD, Tel-Aviv, Israel

**Akira Iguchi**
Terumo Corporation, Tokyo, Japan

**Tomas Kovarnik**
Second Department of Internal Medicine, Charles University, Prague, Czech Republic

**Su-Lin Lee**
EPSRC Center for Interventional and Surgical Sciences, University College London, London, United Kingdom

**Jurgen M.R. Ligthart**
Erasmus MC, Rotterdam, The Netherlands

**John J. Lopez**
Stritch School of Medicine, Loyola University, Maywood, IL, United States

**Josepa Mauri**
University Hospital Germans Trias i Pujol, Badalona, Spain

**Adam Mazurek**
Jagiellonian University, Department of Cardiac & Vascular Diseases, John Paul II Hospital, Krakow, Poland

**Piotr Musialek**
Jagiellonian University, Department of Cardiac & Vascular Diseases, John Paul II Hospital, Krakow, Poland

**Ricardo Ñanculef**
Federico Santa María Technical University, Valparaíso, Chile

**Eric A. Osborn**
Cardiovascular Division, Beth Israel Deaconess Medical Center, Harvard Medical School, Boston, MA, United States

**Lukasz Partyka**
InspireMD, Tel-Aviv, Israel

**Petia Radeva**
Department of Mathematics and Informatics, University of Barcelona, Barcelona; Computer Vision Center, Bellaterra, Spain

**Fernando Ramos**
Department of Mathematics and Informatics, University of Barcelona, Barcelona, Spain

**Josep Rigla**
Department of Mathematics and Informatics, University of Barcelona, Barcelona, Spain

**Juan Rigla**
InspireMD, Tel-Aviv, Israel; Department of Mathematics and Informatics, University of Barcelona, Barcelona, Spain; InspireMD, Boston, MA, United States

**Yuki Sakaguchi**
Terumo Corporation, Tokyo, Japan

**Elias Sanidas**
Department of Cardiology, LAIKO General Hospital, Athens, Greece

**Yusuke Seki**
Terumo Corporation, Tokyo, Japan

**Milan Sonka**
Electrical and Computer Engineering; Iowa Institute for Biomedical Imaging, University of Iowa, Iowa City, IA, United States

**Justyna Stefaniak**
Data Management and Statistical Analysis (DMSA), Krakow, Poland

**Lukasz Tekieli**
Jagiellonian University, Department of Cardiac & Vascular Diseases; Department of Interventional Cardiology, John Paul II Hospital, Krakow, Poland

**Giovanni J. Ughi**
New England Center for Stroke Research, Department of Radiology, University of Massachusetts Medical School, Worcester, MA, United States

**Beatriz Vaquerizo**
Interventional Cardiology Unit, Hospital del Mar, Barcelona, Spain

**Andreas Wahle**
Electrical and Computer Engineering; Iowa Institute for Biomedical Imaging, University of Iowa, Iowa City, IA, United States

**Karen Th. Witberg**
Erasmus MC, Rotterdam, The Netherlands

**Guang-Zhong Yang**
Institute of Medical Robotics, Shanghai Jiao Tong University, Shanghai, China

**Honghai Zhang**
Electrical and Computer Engineering; Iowa Institute for Biomedical Imaging, University of Iowa, Iowa City, IA, United States

**Ling Zhang**
Electrical and Computer Engineering; Iowa Institute for Biomedical Imaging, University of Iowa, Iowa City, IA, United States

**Liang Zhao**
Center for Autonomous Systems, University of Technology Sydney, Ultimo, NSW, Australia

# Editor Biography

**Dr. Balocco Simone** is an associate professor in the Department of Mathematics and Informatics at the University of Barcelona, and a senior researcher at the Computer Vision Center, Bellaterra. He earned his PhD in acoustics from the CREATIS Laboratory, Lyon and in Electronic and Telecommunication from MSD Laboratory, University of Florence (Italy). He conducted a postdoctoral research at the CISTIB Laboratory, Pompeu Fabra University. Dr. Balocco's main research interests include pattern recognition and computer vision methods for the computer-aided detection of clinical pathologies. In particular, his research focuses on ultrasound and magnetic imaging applications and vascular modeling.

# Intravascular Ultrasound: From Acquisition to Advanced Quantitative Analysis

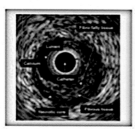
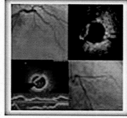
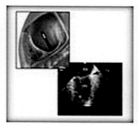
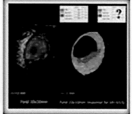

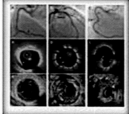
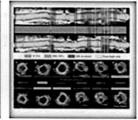

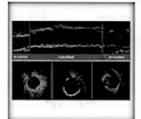
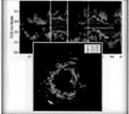

ELSEVIER

# Acknowledgments

I am deeply grateful to "Captain" Juan Rigla, PhD, MD, a great friend who coordinated the clinical part of the manuscript. Without his help, this book would not be possible.

I would like to thank all the contributors of this book for making possible this legacy for the new generation. Thanks go to those who participated and those who wanted but couldn't. Some of them have been working day and night, stealing time from their personal life, and putting aside work, family duties, patients for delivering the best result ever.

My special thanks go to Petia Radeva, the founder of this research line in IVUS, for letting me complete the work that she started so many years ago.

Thanks are due to Fina Mauri and Xavier Carrillo, who were the true motivators of most of this research.

My heartfelt thanks to uncles Carlo Gatta and Francesco Ciompi for being true brothers.

Thanks go to Phillippe Delachartre, who transferred to me his passion for the research in IVUS so many years ago. Thanks are due to Piero Tortoli, Christian Cachard, and Olivier Basset, my mentors now and forever.

I would also like to thank my family, Sara, Giampiero, and Gaetano, for encourage me all along this long journey.

*"...Tell me and I forget, teach me and I may remember, involve me and I will learn...."*

**BENJAMIN FRANKLIN**

# CHAPTER 1

# Introduction

SIMONE BALOCCO[a,b]
[a]Department of Mathematics and Informatics, University of Barcelona, Barcelona, Spain, [b]Computer Vision Center, Bellaterra, Spain

Atherosclerosis, a disease of the vessel walls that causes vessel narrowing and obstruction, is the major cause of cardiovascular diseases such as heart attack or stroke. Intravascular ultrasound (IVUS) is an intraoperative imaging tool that allows visualizing internal vessel structures, quantifying and characterizing coronary plaque, and is useful for diagnostic purposes and image-guided intervention.

The book focuses on imaging, treatment, and computer-assisted technological advances in diagnostic and intraoperative vascular imaging and stenting in IVUS. Such techniques offer increasingly useful information regarding vascular anatomy and function and are poised to have dramatic impact on the diagnosis, analysis, modeling, and treatment of vascular diseases. Computational vision techniques designed to analyze images for modeling, simulating, and visualizing anatomy and medical devices such as stents as well as the assessment of interventional procedures therefore play an important role and are currently receiving significant interest.

The book brings together the advanced knowledge of scientific researchers, medical experts, and industry partners working in the field of IVUS in different anatomical regions.

The book is organized into two sections: the first one includes a clinical perspective of the vascular disease, the current clinical workflow, and the main challenges faced by the research community across anatomical boundaries, including cerebral, coronary, and cardiac interventions. In particular, an introduction for the non-expert readers recalls the basics of this image modality. Special care was taken to describe the relevance of these problems, and to identify the locks that currently hinder their realization.

The second section provides a deep overview of intravascular ultrasound imaging systems analysis. It includes topics of the whole imaging pipeline, ranging from the definition of the clinical problem and image acquisition systems, to the image processing and analysis, including the assisted clinical decision-making procedures and the treatment planning (stent deployment and follow-up). The book will conclude with new research horizons and open questions.

The context of the book is transversal to several disciplines, and specifically it covers several technological fields such as clinical intervention, catheter and IVUS system design, biomechanics, 3D visualization, tissue characterization, segmentation, plaque evolution and rupture analysis, stent deployment, and planning.

For these reasons, the book will be a compendium of the current state of the art in this research field and will be a perfect resource to get updated on the advances of the intravascular ultrasound and stent analysis topics.

Finally, our sincerest thanks go to the authors of all chapters for their dedication to this project and to Elsevier for their support.

**Intravascular Ultrasound.** https://doi.org/10.1016/B978-0-12-818833-0.00001-1

# CHAPTER 2

# Clinical Utility of Intravascular Ultrasound

ELIAS SANIDAS[a] • STÉPHANE CARLIER[b]

[a]Department of Cardiology, LAIKO General Hospital, Athens, Greece, [b]UMONS & CHU Ambroise Paré, Mons, Belgium

## 1. INTRODUCTION

Coronary angiography remains the gold standard imaging method for the detection of coronary artery disease (CAD) and its widespread clinical application has steered patients to a host of beneficial interventional medical therapies. Nonetheless, this approach only provides a two-dimensional image of the contrast-filled arterial lumen and does not visualize the arterial wall where largest atherosclerotic plaques are located. Consequently, angiography often underestimates the degree of intraluminal stenosis and does not gauge the size of the plaque burden itself.[1]

Currently, intravascular ultrasound (IVUS) provides a more detailed assessment of CAD and has emerged as an essential diagnostic tool for understanding coronary lesion morphology, deploying stents, and solving postpercutaneous coronary intervention (PCI) complications.

Historically, the first medical ultrasound application was described in 1953 by Inge Edler and Carl Hertz, who introduced the recording of the motion pattern of cardiac structures along a single sound beam. This technique, known as supersonic reflectoscope, used short supersonic sound pulses that were generated by an electrically excited quartz crystal and delivered to the heart. Part of the sound was reflected back to the quartz crystal and the time difference between the emanation of the sound pulse and the reception of the echo was a measure of the distance between the crystal and the reflecting material. In 1971, the first true IVUS system was designed by Nicolaas Bom and Charles Lancée in Rotterdam. It was conceived as an improved technique for the continuous visualization of cardiac chambers and valves by a catheter with 32 elements with an outer diameter of 3 mm. However, the first transluminal images of human arteries were recorded by Paul Yock in 1988.[2–4]

IVUS is considered a diagnostic imaging method that delivers real-time, high-resolution images of the coronary arteries and provides a precise depiction of the morphology of atherosclerotic plaque.[5] Its role begins with pre-PCI imaging targeted to (1) measure the diameter and the area of the lesion/reference segment, (2) measure the length of the lesion, (3) evaluate the distribution of plaque and presence of calcification, and (4) estimate in vivo plaque composition and burden, identifying plaque characteristics associated with increased vulnerability.[6] In order to further guide percutaneous procedures, IVUS will also be performed poststenting in order to (1) evaluate stent expansion, (2) assess side branch compromise, (3) assess the presence of coronary dissection, and (4) determine the mechanism or stent restenosis or thrombosis (i.e., underexpansion).[7–9]

Notably, 2018 ESC/EACTS Guidelines on myocardial revascularization recommend the use of IVUS to detect stent-related mechanical problems in left main coronary artery (LMCA) as class IIa with level of evidence B.[10] IVUS has also been shown to be an adjunctive imaging technique for the crossing of coronary chronic total occlusions (CTO), the performance of complex aortic, carotid, and peripheral artery endovascular procedures without excluding even vein intervention.

## 2. BASIC PRINCIPLES OF IMAGING ACQUISITION

In summary, the function of IVUS is based on the following general principles:

- conversion of electrical energy into sound waves via piezoelectric crystals;

*Intravascular Ultrasound.* https://doi.org/10.1016/B978-0-12-818833-0.00002-3

- transmission and detection of sound waves reflected by tissues using the same piezoelectric crystals, the transducer, and converting back the received sound waves into an electrical signal;
- amplification and processing of this electrical signal and conversion to an image;
- projection of that image on the device's computer screen from where it can be analyzed or stored.[5]

There are two types of IVUS catheters: the mechanically rotated single-element transducer and the synthetic steered phased array system (Fig. 1). The mechanical catheter has a piezoelectric transducer placed at the edge of a flexible shaft that is rotated and advanced or withdrawn in order to scan the artery within a protective sheath. The systems that have been used lately are high-definition devices running at frequencies between 40 and 60 MHz. The 20-MHz synthetic aperture array catheter has 64 tiny transducers permanently embedded around the circumference of the catheter edge. Cross-sectional images are produced using an electronically phased-array rotating beam forming without any necessary mechanical rotation of the catheter while advanced or withdrawn within the artery.[11,12] The main features of the IVUS catheters are summarized in Table 1.

A gray-scale IVUS image is formed from a codification of the level of echogenicity of the radiofrequency signal that is reflected by the tissues. Signals with low echogenicity are coded as dark gray or black, while highly echogenic signals are coded as light gray or white. The strongest reflection of ultrasound comes from collagen and calcium. The adventitia of the coronary arteries is very rich in collagen and appears as the brightest structure in a noncalcified segment. The external elastic lamina (EEL) is lies between the adventitia and the media, mostly muscular, and typically echolucent (dark). In normal, nonatherosclerotic arteries, the thickness of the media is typically 200 μm. The internal elastic lamina (IEL) separates the media from the most inner structure of the artery, the intima, covered by a single layer of endothelial cells. Intimal thickness increases with age and it is typically 200 μm at 40 years of age producing

the classical three-layer appearance of a normal coronary artery by IVUS. Intimal thickening is the first pathophysiological change related to atherosclerosis. With the accumulation of plaque, intima and IEL tend to merge and the separation from the media is difficult to assess.[13,14]

The definitions of "reference segment" along with the most common measurements using IVUS are presented in Table 2.[15]

## 3. IVUS IN CLINICAL PRACTICE

Atherosclerosis, from the Greek ἀθήρα, *athêra*, meaning "gruel" and σκλήρωσις, sclerosis or "hardening," is by essence a disease of the arterial wall while the lumen will only lately be compromised.[16–18] Thus, an atherosclerotic lesion can evolve during years without any clinical symptom or flow limitation. Previous studies have reported that a coronary angiography performed weeks before an acute myocardial infarction revealed that at the culprit lesion there was only a mild to moderate degree of stenosis in more than half of the patients and as such an angiogram does not provide adequate prognostic information concerning future ischemic events.[19] Such conclusions have given rise to the notion that acute ischemic syndromes are the result of how "vulnerable" an atherosclerotic plaque is to rupture and are less dependent on the degree of luminal stenosis.[20] Several attempts of IVUS signal postprocessing have been reported to detect such vulnerable plaques[21–24]; however, it is seldom used nowadays in a clinical setting. On the other side, numerous studies and metaanalyses compared the clinical outcomes between IVUS-guided and angiography-guided stent implantation with the latest results supporting the utility of IVUS to guide complex PCI procedures, yet remaining underused[25] (Table 3).

### 3.1. IVUS in Percutaneous Transluminal Coronary Angioplasty

IVUS could improve angiographic results by safely upsizing the largest balloon for angioplasty once vessel remodeling was taken into account, as demonstrated in the CLOUT trial.[26] Using balloons sized to the EEL diameter, some advocated aggressive percutaneous transluminal coronary angioplasty (PTCA) instead of systematic stent implantation.[27] Nowadays, with the advent of DES that solved the issues of (1) late lumen loss secondary to negative remodeling post balloon angioplasty,[28] and (2) in-stent restenosis process, such provisional angioplasty strategies based on IVUS or physiological measurements[29] are no more considered.

FIG. 1 Two types of IVUS imaging systems. (A) Mechanical system with a rotating element; (a) cross-sectional image provided by a mechanical system (Atlantis SR Pro Catheter iLab Ultrasound Imaging System). (B) Electronic system with a multielement array; (b) cross-sectional image given by an electronic system (Eagle Eye Gold Catheter S5 System).

TABLE 1
Main Features of Available IVUS Catheters.

| | AltaView (Terumo) | Navifocus WR (Terumo) | AnteOwl WR (Terumo) | Eagle Eye Platinum (Philips) | Eagle Eye Platinum ST (Philips) | Revolution (Philips) | Refinity ST (Philips) | Kodama (ACIST) | OPTICROSS (Boston Scientific) | OPTICROSS HD (Boston Scientific) |
|---|---|---|---|---|---|---|---|---|---|---|
| Frequency | 40-60MHz | 40MHz | 40MHz | 20MHz | 20MHz | 45MHz | 45MHz | 40-60MHz | 40MHz | 60MHz |
| Maximum imaging diameter | 10-32 mm | 10-32 mm | 10-32 mm | 20 mm | 20 mm | 14 mm | 14 mm | 12 mm | 12 mm | 12 mm |
| Diameter at transducer | 2.6 Fr 0.86 mm | 2.5 Fr 0.83 mm | 2.6 Fr 0.86 mm | 3.5 Fr 1.17 mm | 3.5 Fr 1.17 mm | 3.2 Fr 1.07 mm | 3.0 Fr 1 mm | 3.6 Fr 1.2 mm | 2.6 Fr | 2.6 Fr |
| Tip entry profile | - | - | - | 1.5F 0.5mm | 1.5F 0.5mm | 1.7F 0.56mm | 1.9F 0.64mm | 3.2F 1.07mm | 2 Fr 0.66 mm | 2 Fr 0.66 mm |
| Tip-to-transducer length | 22 mm | 9 mm | 8 mm | 10 mm | 2.5 mm | 29 mm | 20.5 mm | 20 mm | 20 mm | 20 mm |
| Working (usable) length | 137 cm | 154 cm | 1358 mm | 150 cm | 150 cm | 135 cm | 135 cm | 142 cm | 135 cm | 135 cm |
| Auto-pullback length | 150 mm | 150 mm | 145 mm | N/A | N/A | 150 mm | 150 mm | 120 mm | 100 mm | 100 mm |
| Wire lumen length | - | - | - | 24 cm | 24 cm | 23 mm | 15 mm | 15 mm | 16 mm | 16 mm |
| Maximum guide wire | 0.014" | 0.014" | 0.014" | 0.014" | 0.014" | 0.014" | 0.014" | 0.014" | 0.014" | 0.014" |
| Minimum guide catheter | 5F (ID ≥ 0.059") | 6F (ID ≥ 0.067") | 6F (ID ≥ 0.070") | 5F (ID ≥ 0.056") | 5F (ID ≥ 0.056") | 6F (ID ≥ 0.064") | 5F (ID ≥ 0.056") | 6F (ID ≥ 0.064") | 5F (ID ≥ 0.058") | 5F (ID ≥ 0.058") |
| Maximum pullback speed | 9 mm/s | 1 mm/s | 9 mm/s | Manual | Manual | 1 mm/s | 1 mm/s | 10 mm/s | 1 mm/s | 1 mm/s |
| Flushing necessary | Yes | Yes | Yes | No | No | Yes | Yes | Yes | Yes | Yes |

**TABLE 2**
Definitions of "Reference Segment" and Measurements by the Use of IVUS.

| Reference segment | Definitions |
| --- | --- |
| Proximal reference | The site with the largest lumen proximal to the stenosis and within the same segment |
| Distal reference | The site with the largest lumen distal to the stenosis and within the same segment |
| Largest reference | The largest site of distal or proximal reference |
| Average reference lumen size | The average value of lumen at distal and proximal reference |
| *Measurements* | |
| Lumen CSA | The area bounded by lumen border |
| Minimum lumen diameter | The shortest diameter of the lumen |
| Maximum lumen diameter | The longest diameter of the lumen |
| Lumen area stenosis | (Reference lumen CSA minus minimum lumen CSA)/reference lumen CSA |
| Plaque plus media CSA | EEL CSA minus lumen CSA |
| Plaque burden | Plaque plus media CSA divided by EEL CSA |
| Superficial calcium | The leading edge of the acoustic shadowing appears within the most shallow 50% of plaque plus media thickness |
| Deep calcium | The leading edge of the acoustic shadowing appears within the deepest 50% of plaque plus media thickness |
| Stent CSA | The area bounded by stent border |

*CSA*, cross-sectional area; *EEL*, external elastic lamina.

On the other side, the importance of optimal lesion preparation before stenting, using rotational atherectomy, cutting balloons, or newer devices, has been identified and IVUS is a very important guidance tool to assess the results of these techniques.[30,31]

## 3.2. IVUS in the Bare Metal Stent Era

Intimal hyperplasia is the major underlying mechanism of bare metal stent (BMS) restenosis. In BMS, percentage of intimal hyperplasia volume averages 30% of stent volume and is consistently greater in diabetics than in nondiabetics. In BMS that do not restenose, initial studies showed that intimal hyperplasia remains stable or regresses slightly after 6 months for a period of up to 2–3 years. Nevertheless, more recent quantitative angiographic data indicate a triphasic BMS luminal response characterized by early restenosis, intermediate-term neointima regression (from 6 months to 3 years), and late re-narrowing (beyond 4 years).[32]

Most of the clinical trials from the BMS era (such as CRUISE,[33] TULIP,[34] DIPOL,[35] AVID[36]) showed a beneficial effect of IVUS in BM stenting, mainly by achieving larger acute lumen dimensions while avoiding increased complications.[32] The MUSIC trial[37] was the first study, followed by a sequence of many others later, that established IVUS criteria for optimal stent implantation. According to the proposed MUSIC criteria, excellent expansion is evident when the minimum lumen area (MLA) in the stent is >90% of the average reference lumen area. All the proposed criteria for IVUS optimization used in different studies have relied on distal reference or on mean reference vessel for stent or post-dilatation balloon sizing. However, this fact reduces the potential to optimally increase the lumen size particularly in long lesions with overlapping stents and in vessels with distal tapering.

A metaanalysis of randomized trials comparing IVUS with angiographic-guided BMS implantation (n = 2193 patients) showed that IVUS guidance was associated with a significantly lower rate of angiographic restenosis, repeat revascularization, and overall major adverse cardiac events (MACE), but had no significant effect on myocardial infarction.[38] Conversely, the results of the OPTICUS trial did not show a significant difference between the angiographic- and IVUS-guided groups not supporting the routine use of ultrasound guidance for coronary stenting.[39] Likewise, a large metaanalysis that included five randomized trials and four registries with a total of 2972 patients found that the primary endpoint (occurrence of death or nonfatal myocardial infarction) was similar for both strategies at 6 months. Pooled data of individual cardiac endpoints showed a 38% reduced probability of target lesion revascularization (TLR) in favor of IVUS-guided stenting, while death, nonfatal myocardial infarction, or coronary artery bypass graft were equally distributed in both groups.[40]

**TABLE 3**
Clinical End Points of Studies and Metaanalyses That Compared IVUS With Angiography Guidance in PCI.

| Study | Year | Patients | Follow-up | RESTENOSIS RATE | | | MACE | | | MYOCARDIAL INFARCTION | | | MORTALITY RATE | | |
|---|---|---|---|---|---|---|---|---|---|---|---|---|---|---|---|
| | | | | IVUS group | Angio group | P-Value | IVUS group | Angio group | P-Value | IVUS group | Angio group | P-Value | IVUS group | Angio group | P-Value |
| Parise et al. | 2011 | 2193 | 6–30 months | 22% | 29% | .02 | 19% | 23% | .03 | 35% | 44% | .51 | 25% | 17% | .18 |
| Mudra et al. | 2001 | 550 | 6–12 months | 24.5% | 22.8% | .68 | NA | NA | NA | NA | NA | NS | NA | NA | .12 |
| Casella et al. | 2003 | 2972 | 6 months | 0.6% | 1% | .2 | 15% | 18.7% | .03 | 4.1% | 3.7% | .5 | 0.6% | 0.6% | 1 |
| Zhang et al. | 2012 | 19,619 | 20.7 ± 11.5 months | NA | NA | <.001 | NA | NA | .008 | NA | NA | .126 | NA | NA | <.001 |
| Hong et al. | 2015 | 1323 | 12 months | NA | NA | NA | 2.9% | 5.8% | .007 | NA | NA | NA | 0.4% | 0.7% | .48 |
| Steinvil et al. | 2016 | 31,283 | 9–48 months | NA | NA | <.001 | NA | NA | <.001 | NA | NA | <.001 | NA | NA | <.001 |
| Elgendy et al. | 2016 | 3192 | 12–24 months | 0.6% | 1.3% | .04 | 6.5% | 10.3% | <.0001 | 0.8% | 1.5% | .06 | 0.5% | 1.2% | .05 |
| Bavishi et al. | 2017 | 3276 | 1.4 ± 0.5 years | 0.6% | 1.3% | .15 | 6.5% | 10.5% | .0001 | 2% | 2.4% | .65 | 0.5% | 1.2% | .09 |
| Shin et al. | 2016 | 1170 | 12 months | 0.3% | 0.5% | .32 | 0.4% | 1.2% | .04 | 0 | 0.4% | .02 | 0.3% | 0.7% | .134 |
| Witzenbichler et al. | 2014 | 8583 | 12 months | 0.6% | 1% | .003 | 3.1% | 4.7% | .002 | 2.5% | 3.7% | .004 | 0.8% | 1.2% | .12 |
| Maehara et al. | 2018 | 8586 | 24 months | 0.55% | 1.16% | .003 | 4.9% | 7.5% | .003 | 3.5% | 5.6% | .0006 | 1.7% | 2.4% | .03 |
| Zhang et al. | 2018 | 1448 | 12 months | 1.5% | 2.9% | .07 | 2.9% | 5.4% | .02 | 1% | 1.5% | .34 | 0.7% | 1.4 | .19 |

*NA*, nonavailable; *NS*, no significant.

## 3.3. IVUS in the Drug-Eluting Stent Era

Drug-eluting stent (DES) implantation is commonly associated with very few clinical events. IVUS predictors associated with PCI failures and increased adverse outcomes with DES include stent underexpansion, nonuniform strut distribution, edge-related problems such as residual reference disease (geographic miss) and dissections, as well as acute and, especially, late incomplete stent apposition (malapposition).[9,41–45]

Stent underexpansion results from poor expansion during implantation rather than from chronic stent recoil and may be undetectable angiographically in many cases; suspicion may be raised in an area of fluoroscopically underexpanded stent struts (compared with the rest of struts) in the context of a calcified lesion or an inability to fully expand the balloon inside the stent. Nevertheless, the use of IVUS can be instrumental to detect underexpansion; despite good apposition of the stent struts to the vessel wall, the underexpanded site would be evident by a stent cross-sectional area significantly smaller than the vessel cross-sectional area in the same site, smaller than the stent cross-sectional area in other sites, and smaller than the reference lumen area. According to proposed strict criteria, excellent expansion is evident when the MLA in the stent is $\geq$90% of the average reference lumen area.[44]

A condition that needs to be differentiated from underexpansion is stent malapposition; unlike underexpansion, there are stent struts not apposed to the vessel wall (i.e., space occupied by blood can be detected between the stent struts and the arterial intima). Malapposition cannot be judged angiographically (except in very few extreme cases), typically occurs with use of undersized stents or in arteries that have significant tortuosity and fluctuations of reference arterial lumen diameter within the treated segment and is thought to predispose to stent thrombosis. However, no association was found between early or late incomplete stent apposition and stent thrombosis in 1580 patients enrolled in IVUS substudies of various TAXUS clinical trials. Because both malapposition and underexpansion affect selected regions of a stent, it is entirely possible that they coexist in two separate sites of the same stent (i.e., proximal struts can be malapposed owing to large and tortuous proximal reference sites, whereas the mid stent area at the original lesion site can be underexpanded).[44,46]

Whether IVUS compared to angiography guidance reduces stent thrombosis and improves clinical outcomes associated with DES treatment has been investigated over the years. In a metaanalysis of 11 clinical studies ($n = 19{,}619$), IVUS-guided DES implantation as compared with angiography-guidance alone was associated with a reduced incidence of death, MACE, and stent thrombosis.[47] The IVUS-XPL study demonstrated that IVUS-guided DES implantation resulted in lower rate of MACE (2.8%, HR, 0.47 [95% CI, 0.27–0.82]; $P = .007$, per protocol analysis) compared to angiography-guided stent implantation (5.9%) at 1-year follow-up among 1323 patients with long coronary lesions, primarily due to lower risk of TLR. In the post hoc analysis of the patients of the IVUS-guided group, those who did not meet the IVUS criteria had a significantly higher incidence of MACE compared with those meeting the IVUS criteria for stent optimization (4.6% vs 1.5%; HR 0.31 [95% CI, 0.11–0.86], $P = .02$). These data support strongly IVUS guidance in such lesions.[48]

An updated metaanalysis of 7 randomized control trials and 18 observational studies confirmed the above data and found that IVUS-guided PCI was associated with better clinical outcomes including MACE, mortality, stent thrombosis, TLR, and target vessel revascularization than angiography-guided DES implantation. However, in a separate analysis that included only the randomized control trials, the observed benefit for MACE was driven only by reduced rates of revascularizations.[49] Similarly, a metaanalysis of seven randomized trials ($n = 3192$) showed that IVUS-guided PCI was not inferior to angiography-guided PCI in reducing the risk of MACE (6.5% vs 10.3%) mainly due to the reduction in the risk of TLR (4.1% vs 6.6%). The risk of cardiovascular mortality (0.5% vs 1.2%) and stent thrombosis (0.6% vs 1.3%) were also lower in the IVUS-guided group.[50]

ADAPT-DES was a prospective, multicenter, real-world study of 8583 consecutive patients at 11 international centers undergoing DES implantation that investigated the frequency, timing, and correlation between stent thrombosis and adverse clinical outcomes post-PCI. During the index procedure, IVUS was used in 3349 patients. IVUS performance resulted in longer stent length and larger stent size without increasing periprocedural myocardial infarction. It was shown that IVUS guidance led to less stent thrombosis (0.6% vs 1%) beginning at the time of implantation, as well as fewer myocardial infarction (2.5% vs 3.7%) within the first year.[51] The benefits of IVUS guidance further increased with long-term follow-up of up to 2 years.[52]

Currently, the ULTIMATE trial that included 1448 all-comer patients who were randomly assigned (1:1 ratio) to either IVUS guidance or angiography guidance before DES implantation indicated a significant reduction in MACE at 12-month follow-up in IVUS-guided compared to angiography-guided group (2.9% vs 5.4%, respectively, HR: 0.530; 95% CI: 0.312–0.901; $P=.019$). In the IVUS group, TVF was 1.6% for patients with successful procedures versus 4.4% in patients who failed to achieve all optimal criteria (HR: 0.349; 95% CI: 0.135–0.898; $P=.029$), namely MLA in the stented segment $>5.0\,mm^2$ or >90% of the MLA at the distal reference segments, plaque burden 5 mm proximal or distal to the stent edge is <50%, and no edge dissection involving the media with a length $>3\,mm$.[53] In the all-comers, open-label, single-arm SYNTAX II study that investigated the impact of a state-of-the-art PCI strategy on clinical outcomes in patients with triple vessel disease, final minimal stent area measured by IVUS was available in 819 lesions in 367 patients (53% of lesions, 81% of patients). The single postprocedural MSA value that best separated lesions with TLR from those with no TLR was $5.2\,mm^2$.[54]

A large registry of more than 6000 patients undergoing PCI with DES for complex lesions (defined as bifurcation, chronic total occlusion, left main disease, long lesion, multivessel PCI, multiple stent implantation, in-stent restenosis, or heavily calcified lesion) reported recently that IVUS guidance lead to a significantly lower risk of cardiac death during 64 months of median follow-up (10.2% vs 16.9% with angiography-guided PCI; HR: 0.573; 95% CI: 0.460–0.714; $P<.001$).[55] With IVUS, the implanted stents had a significantly larger mean diameter ($3.2\pm0.4$ vs $3.0\pm0.4$; $P<.001$) and were more frequently postdilatated (49.0% vs 17.9%; $P<.001$).

## 4. IVUS APPLICATIONS

Gray-scale IVUS offers only a limited characterization of the atherosclerotic plaque composition and is unable to detect the specific histomorphological features that are associated with the rupture of vulnerable plaques, namely a large and lipid-rich necrotic core; a thin and inflamed fibrous cap, rich in macrophages, that covers the necrotic core; and neovascularization.[56,57] For these reasons, novel IVUS techniques such as integrated backscatter IVUS (IB-IVUS), VH-IVUS, iMap-IVUS, near-infrared spectroscopy-IVUS (NIRS-IVUS), and contrast-enhanced IVUS (CE-IVUS) have been developed.[5,11]

### 4.1. IB-IVUS

A tissue classification scheme based only on the analysis of the integrated backscattered (IB) signal, using a simple surface scanner on carotid samples was primarily described in 2001.[58] This methodology was developed with the integrated, rotating, 40-MHz IVUS catheter from Boston Scientific (Fremont, California, United

States). IB-IVUS was applied to 18 samples of coronary artery and the results were compared with the corresponding histological findings. The resulting IB-IVUS values were divided into five categories so that coded color maps could be constructed: thrombus, intimal hyperplasia or lipid core, fibrous tissue, mixed lesions, and calcification. The initial comparisons between angiography and IB-IVUS showed that the angioscopically colored surface of the plaque reflected the thickness of the fibrous cap more than the size of the lipid core.[59] Initially developed with the Boston Scientific Clearview system, IB-IVUS (IB-IVUS, YD Co, Ltd., Nara, Japan) is now used with the VISIWAVE platform from Terumo (Tokyo, Japan). With the ViewIT 40-MHz catheter (Terumo, Tokyo, Japan), a comparison of IB-IVUS with histopathology shows that the sensitivity in the classification of calcification, fibrous tissue, and lipids was 90%, 84%, and 90%, while specificity was 97%, 96%, and 86%, respectively.[60]

### 4.2. VH-IVUS

VH-IVUS allows tissue characterization of vascular lesions and is based on the spectral analysis of the primary raw backscattered ultrasound wave (radiofrequency-based signal). This method has an estimated axial resolution (based on the resolution of the 20-MHz IVUS catheter) of approximately $200\,\mu m$ (Table 4). The spectral signatures of four tissue types (fibrous tissue, fibrofatty tissue, necrotic core, and dense calcium) were determined in vitro.[21] These signatures are programmed in the software of the IVUS console or on a stand-alone software package for off-line analysis of patient data. Radiofrequency-IVUS plaque components are color coded as dense calcium (white), necrotic core (red), fibrofatty (light green), and fibrous tissue (dark green).[61,62] Correlation between gray-scale and VH-IVUS imaging is shown in Fig. 2.

Ex vivo validation of VH images directly with the histopathology sections provided accuracies of up to 97%. Independent studies have demonstrated in vivo a relatively high level of accuracy and reproducibility of VH-IVUS in human arteries utilizing directional coronary atherectomy specimens, yielding predictive accuracies of up to 95% in nonacute coronary syndrome (ACS) patients.[61] Of note, such a sequential assessment is impacted by precise co-registration of the histologic tissue sample and the imaging. In adult atherosclerosis-prone minipigs, no correlation was observed between the size of the necrotic core determined by VH-IVUS and histology.[63]

The PROSPECT trial tried to assess the natural history of atherosclerosis by studying 697 patients with ACS

**TABLE 4**
**Detection and Technical Parameters of IVUS and VH-IVUS.**

|  | IVUS | VH-IVUS |
|---|---|---|
| *Detection* | | |
| Lipid/necrotic core | + | ++ |
| Fibrous cap | + | +++ |
| Thrombus | + | No |
| Calcium | +++ | +++ |
| Plaque rupture | ++ | No |
| Attenuated plaque | +++ | No |
| TCFA (thin-cap fibroatheroma) | No | ++ |
| Dissection | ++ | No |
| Stent expansion/apposition | ++ | No |
| Stent strut coverage | + | + |
| *Technical parameters* | | |
| Frequency (MHz) | 20–45 | 20–45 |
| Frame rate | 10–30 | 10–30 |
| Pullback speed (mm/s) | 0.5–1 | 0.5–1 |
| Axial resolution (μm) | 70–200 | 70–200 |
| Tissue penetration (mm) | >5 | >5 |
| Ease of use | +++ | ++ |
| Need for contrast | No | No |

*Notes*: +: Limited possibilities only; ++: well suited; +++: excellent.

after successful PCI of a culprit lesion under optimal medical therapy using angiography plus three-vessel imaging including gray scale and VH-IVUS. In patients with ACS, both culprit and nonculprit lesions were equally likely to spur subsequent adverse events such as cardiac death, cardiac arrest, myocardial infarction, or rehospitalization due to unstable or progressive angina over 3 years. Independent predictors of a future cardiovascular event were plaques classified as VH-TCFAs (fibroatheroma without evidence of a fibrous cap: >10% confluent NC with >30° NC abutting the lumen in at least three consecutive frames) with a plaque burden >70% and an MLA <4 mm$^2$.[64]

Similarly, the VIVA study was a prospective analysis of 170 patients with stable angina or ACS who underwent three-vessel VH-IVUS before and after PCI. At a median of 1.7 years, 19 lesions (13 nonculprit and 6 culprit) resulted in major adverse cardiac events (MACE including death, myocardial infarction, unplanned revascularization). Nonculprit lesion factors associated with nonrestenotic MACE were VH-IVUS thin-capped fibroatheroma (TCFA), plaque burden >70%, and MLA <4 mm$^2$ suggesting that VH-IVUS can identify plaques at an increased risk of subsequent events.[65] More recently, after a mean follow-up of 51 ± 6 months of 86 patients with 89 intermediate lesions defined as 30%–70% stenosis in coronary angiography, MACE were found to be significantly related to angiographic diameter stenosis, fibrofatty area (FFA) but not necrotic core, IVUS plaque burden ≥70%, and area stenosis

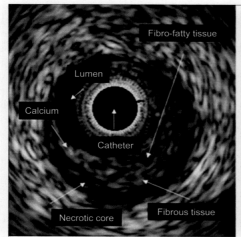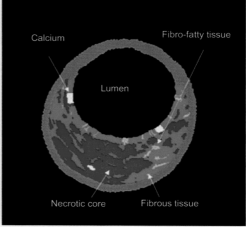

FIG. 2 Correlation between gray-scale IVUS and VH-IVUS imaging. These two cross-sectional frames depict the same arterial location and allow visualization of a significant eccentric atherosclerotic plaque. Gray-scale IVUS *(left)* can easily identify lumen and plaque borders, but VH-IVUS *(right)* provides additional information regarding the compositional plaque characteristics.

≥50%. In a multivariable analysis, only FFA was independently associated with the occurrence of MACE (HR 1.36, 95% CI 1.05–1.77, $P=.019$).[66] The differential predictive value of the IVUS-VH-derived necrotic core or fibrofatty components among different studies illustrates the challenges of IVUS-derived tissue characterization, from the reliability of the training dataset to the complexity of the recognition algorithms[67] with underestimated issues initially such as the lack of enough ultrasound signal for any spectral analysis behind calcified lesions.[68]

## 4.3. iMap-IVUS

With this software, results are presented in a way similar to the VH-IVUS system. However, there are differences: it is based on a full spectrum analysis and the applied color scheme shows (i) fibrous tissue in light green, (ii) lipid tissue in yellow, (iii) necrotic core in pink, and (iv) calcium in blue.[23] Furthermore, the applied IVUS catheter is a 40-MHz rotating single-element catheter instead of the 20-MHz mechanical one used in VH-IVUS. Ex vivo validation demonstrated accuracies at the highest level of confidence (97%, 98%, 95%, and 98% for necrotic, lipid, fibrotic, and calcified regions, respectively).[69,70] In a study of 87 patients, iMap analysis showed that the ACS lesions had larger lipidic and necrotic components compared to non-ACS lesions.[71] Among 63 patients with ST-segment myocardial infarction (STEMI), iMap-IVUS detected a higher percentage of necrotic tissue in culprit lesions that remained high after 10 months whereas the proportion of lipidic tissue decreased.[72] It can also predict slow flow, a serious complication of PCI that is correlated with poor prognosis.[73] Finally, iMap evaluates the neointimal tissue after stent placement providing additional insights.[74]

## 4.4. NIRS-IVUS

Infrared region is the most reliable for the documentation of complex molecules because each bond vibration contributes to a fingerprint of the molecule. Identification of a lipid core plaque (LCP) is based on the distinction of cholesterol spectral features differentiating cholesterol from the other chemicals present and especially collagen. The NIRS-IVUS system provides real-time chemical measurements in the coronaries. It includes a console, a pullback motor unit, a rotation device, and a catheter that automatically scans the artery like IVUS. Spectra are processed by a specific algorithm trained in vitro[75] and displayed as a chemical image of lipid-rich plaque probability (called the "chemogram"), which is depicted as yellow (Fig. 3). The system acquires

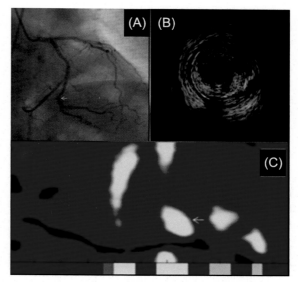

FIG. 3 Multimodality assessment in the catheterization laboratory. An intermediate angiographic lesion located at the distal LCX (A). The MLA measured by IVUS was 4.0 mm² and the plaque burden 63% (B). The related chemogram shows *yellow areas* indicating lipid core plaque (C). Physiologic lesion assessment after intravenous administration of adenosine demonstrated a fractional flow reserve (FFR) of 0.80. The lesion was finally treated with a drug-eluting stent (DES).

1000 measurements/12.5 mm and each measurement interrogates 1–2 mm² of lumen. The available LipiScan IVUS (InfraReDx, Inc., Burlington, Massachusetts, United States) combines a 40-MHz rotational IVUS imaging system along with the NIRS advanced technology. This allows a complete visualization of coronary structure and plaque morphology together with a detailed chemical map of the vessel for the simultaneous detection and localization of LCP.[61]

NIRS-IVUS is highly accurate in detecting LCP in human coronary arteries and show the existence and distribution of necrotic core, but not the amount or the fibrous cap thickness. Increasing evidence is linking LCP to vulnerable plaque, lesions at risk for embolization, and stent thrombosis. As part of a continuing goal to understand the linkage between NIR signals indicating the presence of a lipid-rich plaque and subsequent coronary events, the COLOR registry showed that the absence of a lipid-rich plaque is associated with good prognosis. In particular, PCI performed in lesions with large lipid core was correlated with a 50% risk of periprocedural myocardial infarction compared with only a 4.2% risk in lesions without large lipid core.[76]

The CANARY study aimed to evaluate criteria for defining LCP that is at high risk of rupturing during standard-of-care therapy and causing intraprocedural complications such as distal embolization. According to this study, plaques responsible for periprocedural myonecrosis were lipid rich and had a large plaque burden and a small MLA. However, nonlipid rich plaques could evoke a substantial proportion of myocardial infarction.[77]

Finally, the LRP study is a prospective, multicenter trial that was designed to investigate the correlation between LCP and the occurrence of MACE. It included 1563 patients with suspected CAD (46.3% stable angina) that underwent angiography and possibly PCI for an index event. All the patients were evaluated with IVUS-NIRS to assess the vessel structure and the plaque composition and were followed for 2 years. According to this trial, the risk for a nonculprit or unstented MACE event was 18% higher with each 100 unit increase in maxLCBI4 mm (maximum lipid burden in any 4 mm subsegment). However, the risk in case of a vulnerable coronary segment was 45% higher with each 100 unit increase in maxLCBI4 mm. Of note, the risk in a coronary segment with a maxLCBI4 mm >400 was 422% higher than a segment with a lesser maxLCBI4 mm.[78]

### 4.5. Contrast-Enhanced IVUS

Neovascularization in an atherosclerotic plaque has been linked to plaque growth and instability and contrast-enhanced IVUS (CE-IVUS) is a proposed method for the detection of vasa vasorum (VV), the microvessels that nourish the vessel walls. CE-IVUS is based on the infusion of contrast microbubbles, which can cause an increase in the echogenicity of selected regions on the IVUS images that include atheromatous plaques. It is able to record qualitatively and quantitatively the flow (presence) of microbubbles in human atherosclerotic plaques, mainly within the microvessels and neovasculature using specially developed software for this purpose (Fig. 4).[79–82]

The perivascular network was examined using CE-IVUS in an animal study that aimed to detect blood flow into the coronary lumen and perivascular flow. A statistically significant enhancement was found in the echogenicity of the total perivascular space (adventitial region and perivascular vessels), as indicated by an increase in gray level intensity after microbubble injection.[83] A recently published study also showed that CE-IVUS images could detect aortic wall neovascularization in rabbits being in agreement with histological data.[84]

The implementation of this method in ACS patients demonstrated that CE-IVUS images could shed light on the neovascularization of the adventitia and within the atherosclerotic plaque region. Hence, CE-IVUS can provide important information concerning the presence of a vulnerable plaque and cardiovascular risk stratification.[81]

### 5. IVUS ASSESSMENT OF PLAQUE PROGRESSION

Ultrasound waves from IVUS catheters can travel deep enough to image completely the thickened atherosclerotic vessel wall and study atherosclerosis progression.[1] Several studies have indicated the value of IVUS in evaluating plaque volume regression over time using different treatment. The IVUS-derived rate of progression of atherosclerotic burden is a surrogate end point that could reflect the beneficial clinical impact of the investigated therapies.

Early reports have described a reduction in the lipoid components and an increase in fibrous tissue after statin therapy.[85] The REVERSAL trial was the first double-blind randomized multicenter study that demonstrated a difference in the effects of two statins (atorvastatin vs pravastatin) administered for 18 months.[86] The IVUS-derived change in atheroma volume showed a significantly lower coronary plaque progression rate in the atorvastatin group. The ESTABLISH trial that included ACS patients showed that aggressive lipid-lowering atorvastatin therapy decreased significantly the plaque volume after 6-month follow-up, positively correlated with the LDL level.[87] The most important changes in IVUS measurements (progression and regression) were seen in the ASTEROID trial. Among patients with an ACS or stable CAD who received intensive therapy with rosuvastatin, LDL level decreased by 53% while mean percentage atheroma volume for the entire vessel was lower by 1% ± 3%.[88]

IVUS plaque regression has been also investigated with other therapies such as ezetimibe and PCSK9 inhibitors. The PRECISE-IVUS study revealed a greater plaque regression combining ezetimibe with statins probably due to the most aggressive lipid-lowering effect because of the inhibition of cholesterol absorption by ezetimibe.[89]

Additionally, the GLAGOV trial showed that among 868 patients with CAD, the addition of evolocumab to statin therapy resulted in a significant decrease in LDL-C levels (93.0 vs 36.6 mg/dL) and the primary efficacy parameter, the nominal change in percent atheroma

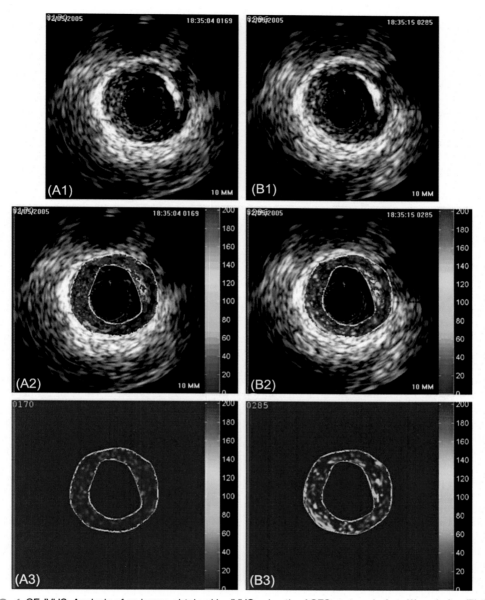

FIG. 4 CE-IVUS. Analysis of an image obtained by IVUS using the ACES system, before (A) and after (B) the infusion of contrast microbubbles. (1) Initial IVUS image; (2) IVUS image from the region of the plaque under analysis; and (3) result of processing without the original image.

volume (PAV), increased 0.05% with placebo and decreased 0.95% with evolocumab (difference, −1.0% [95% CI, −1.8% to −0.64%]; $P < .001$). This PCSK9 inhibitor by further reducing LDL cholesterol induced plaque regression in a greater percentage of patients than placebo (64% vs 47%; $P < .001$ for PAV).[90]

Nevertheless, in the GLAGOV RF-IVUS substudy, there was no significant change in plaque composition such as calcium, fibrous, fibrofatty, or necrotic core tissues between the two groups.[91] Last but not least, the ODYSSEY J-IVUS is an ongoing project that was designed to investigate the effect of alirocumab compared to statin monotherapy on coronary atherosclerotic plaque volume in Japanese patients with a recent ACS and hypercholesterolemia.[92]

## 6. IVUS IN COMPLEX CORONARY LESIONS

Since the introduction of the second-generation DES, the rate of PCI failure has markedly decreased. Nevertheless, the issue of adequate stent implantation becomes even more important, especially in regard to complex coronary lesions such as LMCA disease, multivessel disease, long lesions, coronary calcification, and CTO.[61]

In patients with complex lesions (i.e., bifurcations, long lesions, CTOs, or small vessels), treated exclusively with DES, the use of IVUS in the AVIO trial demonstrated a benefit in MLA after stenting when compared to angiography alone. Nevertheless, no statistically significant difference was found in MACE up to 24 months.[93] The efficacy and safety of IVUS-guided PCI in patients with complex coronary lesions, treated with DES, was also investigated in another metaanalysis of eight randomized trials ($n = 3276$) that indicated the superiority of IVUS-guided interventions. Patients undergoing IVUS-guided PCI had significantly lower risk for MACE, TLR, and target vessel revascularization. Nonetheless, there were no significant differences for stent thrombosis, cardiovascular death, or all-cause mortality between the two groups.[94] A metaanalysis of three randomized trials included 2345 patients with long lesions or CTOs also demonstrated that in a year postprocedure MACE had occurred in 0.4% of the patients who underwent IVUS-guided new generation DES implantation compared to 1.2% of those who underwent angiography-guided implantation.[95]

A recently published study that included 6005 patients with at least one complex lesion (bifurcation, CTO, LMCA disease, long lesion, multivessel PCI, multiple stent implantation, in-stent restenosis, or heavily calcified lesion) confirmed the beneficial effects of IVUS-guided PCI compared with angiography-guided PCI. Patients with IVUS guidance had a significantly larger mean stent diameter, longer stent length, and more frequent use of postdilatation compared to those with angiography guidance. Moreover, in IVUS-guided group, the risk of all-cause mortality, myocardial infarction, stent thrombosis, and MACE was significantly lower. Of note, the greatest benefit of IVUS guidance was observed in patients with LMCA disease, although favorable outcomes were identified in most complex coronary lesions.[55]

### 6.1. Left Main Coronary Artery Lesions

Left main coronary artery (LMCA) lesions have proven to be notoriously difficult to be accurately evaluated by angiography alone. Angiographic appraisal of left main disease correlates very poorly with IVUS and fractional flow reserve (FFR) determinations of lesion severity. This is related to high intra- and interobserver variability as well as the angiographic underestimation of left main dimensions. Moreover, the extent of left main bifurcation plaque burden by IVUS influences PCI outcome and, in general, PCI of distal left main bifurcation lesions are related to poorer prognosis. IVUS is very useful in distinguishing significant from insignificant left main disease, depicting the distribution of plaque and planning the appropriate treatment strategy.[61] Indeed, among 115 patients with angiographically intermediate LMCA stenosis, fewer than half had significant stenoses in IVUS evaluation, especially for lesions located at the left main ostium.[96]

Several studies have differed over the years regarding the cutoff value of the IVUS-derived MLA that is suggestive of a significant LMCA stenosis. Among 121 patients with angiographically normal LMCAs that were evaluated with IVUS, the deferral of revascularization in case of an MLA $\geq 7.5$ mm$^2$ seemed to be safe. Furthermore, based on clinical long-term outcomes, the best cutoff of the MLA for performing or deferring revascularization was 9.6 mm$^2$.[97] Another study of 55 patients with an angiographically ambiguous LMCA stenosis showed that an MLA of 5.9 mm$^2$ had 93% sensitivity and 95% specificity for determining a significant LMCA stenosis as defined by a fractional flow reserve (FFR) <0.75.[98] Korean authors have suggested that in isolated LM disease, an IVUS-derived MLA <4.5–4.8 mm$^2$ in their populations was useful for predicting an FFR <0.80.[99,100] However, differences based on ethnicity must be taken into account: a study compared coronary LM lesions between 99 white North American and 99 Asian patients matched on basis of age, gender, and diabetes mellitus and demonstrated that Asian patients had a significantly smaller LM-MLA ($5.2 \pm 1.8$ mm$^2$ vs $6.2 \pm 1.4$ mm$^2$; $P < .0001$).[101]

Given the unique prognostic implications of LMCA disease, the European Bifurcation Club recommends using a threshold MLA cutoff of 6 mm$^2$ to indicate an LMCA lesion that should be treated with revascularization in a European population.[102] Data from patients with intermediate LMCA lesion showed similar long-term results when FFR > 0.8 or MLA > 6 mm$^2$ were used to defer revascularization compared to patients with FFR < 0.8 treated with revascularization.[103]

In 2011, the multicentric LITRO study validated the MLA in 354 patients included in 22 centers. LMCA lesion was revascularized in 91% (152 of 168) of them with MLA < 6 mm$^2$ while it was deferred in 96% (179 of 186) with MLA $\geq$ 6 mm$^2$. After 2-year follow-up, cardiac death-free survival rate was 95% in the revascularized group versus 98% in the deferred group whereas event-free survival rate was 81% versus 87%, respectively.[104]

In the MAIN-COMPARE registry, 975 patients underwent unprotected left main stenting; of those, 756 had IVUS guidance and 219 did not. In particular, the comparison between 145 equivalent matched groups of patients who received DES showed that IVUS guidance in left main PCI was associated with reduced long-term myocardial infarction and mortality. According to the same data, the optimal minimum stent area in left main lesions to prevent TLR was 8.7 mm$^2$.[105]

In the EXCEL IVUS substudy, the post-PCI minimum stent area was 9.9 $\pm$ 2.3 mm$^2$ and was strongly correlated with adverse events such as death, myocardial infarction, and stent thrombosis during the 3-year follow-up.[106] IVUS was also performed pre-PCI in 270 (47%) and post-PCI in 430 (74%) of 580 PCI-treated patients with LMCA lesions in the NOBLE trial which suggested that coronary artery bypass grafting (CABG) might be better than PCI for the treatment of LMCA disease.[107] However, there are numerous ongoing trials that are expected to shed light on the management of patients with LMCA.[108]

## 6.2. Other Coronary Lesions

The first report showing that an IVUS-derived minimal lumen area (MLA) >4 mm$^2$ was correlated with a coronary flow reserve $\geq$2 was published in 1998.[109] After 1 year, an MLA < 4 mm$^2$ was found to be associated with ischemia on scintigraphy.[110] Currently, available data indicate that the MLA threshold suggestive of myocardial ischemia range from 2.1 to 4.4 mm$^2$. Nevertheless, this index has high negative predictive value and positive predictive value indicating that it is surely safe to defer the intervention of a lesion with MLA > 4 mm$^2$. However, using the cutoff of MLA < 4 mm$^2$ to perform a PCI could lead to treatment of 50% of lesions without ischemia.[111] Among 300 patients with a deferred revascularization, because MLA > 4 mm$^2$, the MACE rate was only 4.4% and the TLR rate was 2.8% at 1-year follow-up.[112]

On the other hand, the FFR, which is an invasive physiological index that predicts coronary stenosis-induced ischemia, has been extensively validated against noninvasive ischemia testing and discriminates ischemic and nonischemic lesions with an accuracy of 95%.[61] According to a study that evaluated either FFR or MLA in 167 consecutive patients with intermediate coronary lesions, a total of 94 lesions were detected in the IVUS guidance arm while 83 lesions in the FFR group. The IVUS-guided group underwent revascularization therapy significantly more often, but at 1-year follow-up, no significant difference was found in MACE rates between the two groups.[113]

Moreover, in a total of 692 consecutive patients with 784 coronary lesions that were evaluated by IVUS and FFR before intervention, the best cutoff value of MLA for predicting FFR < 0.8 was 2.4 mm$^2$ with sensitivity 84% and specificity 63%. In the subgroup analysis, the MLA cutoff was 2.4 mm$^2$ for the left anterior descending coronary artery, 1.6 mm$^2$ for the left circumflex coronary artery, and 2.4 mm$^2$ for the right coronary artery.[114]

The FLAVOR study is an international, multicenter, prospective randomized ongoing clinical trial where a total of 1700 patients with intermediate coronary artery stenosis will be randomized 1:1 to receive either FFR-guided or IVUS-guided stenting. This study has been designed to evaluate the safety and the efficacy of two strategies comparing the rate of all-cause mortality, myocardial infarction, and revascularization after 24-month follow-up.[115]

## 6.3. Calcified Lesions

Calcium is a powerful reflector of ultrasound since little beam penetrates calcium and, thus, it casts a shadow over deeper arterial structures. By the use of IVUS, calcium is illustrated as echodense plaque that is brighter than the reference adventitia with shadowing. Additionally, multiple reflections are produced by the oscillation of ultrasound between the transducer and calcium causing concentric arcs at reproducible distances. Calcium can be assessed (1) quantitatively by measuring the arc in degrees and the length; (2) semiquantitatively classifying calcium as absent or subtending 1, 2, 3, or 4 quadrants; and (3) qualitatively based on its location (lesion vs reference or superficial vs deep). Notably, IVUS is considered more accurate compared to angiography alone for detecting the coronary calcification. In a study of 1155 native vessel target lesions, IVUS detected calcium in 73% of lesions, significantly more often than standard angiography (38%).[7,116,117] Coronary calcium is associated with atherosclerotic plaque growth. In 101 IVUS-detected ruptured plaques compared to 101 computer-matched control plaques without rupture, there was quantitatively less calcium, especially superficial calcium, but a larger number of small calcium deposits, mainly deep calcium deposits.[118] The rupture of a thin fibrous cap atheroma might also be related to minute (10-μm-diameter) cellular-level microcalcifications in the cap, a hypothesis experimentally confirmed on autopsy specimens using in vitro imaging techniques with a resolution exceeding intravascular ultrasound and optical coherence tomography.[119] Atheroma that are actively undergoing calcification are most likely to cause MACE. Recently, molecular PET/CT imaging with ionic 18F fluoride has been developed to identify such lesions.[120]

### 6.4. Coronary Chronic Total Occlusion

In coronary chronic total occlusion (CTO) PCI, IVUS plays an important role in guiding the recanalization of the CTO and for stent optimization.[121] Initial reports of IVUS guidance after CTO lesion crossing to provide information about lesion length and morphology and identify a safe intraluminal landing zone for stenting are more than 20 years old,[122] yet this approach remains considered as a niche indication performed by very few interventionalists and only mastered by our Japanese colleagues.[123,124] An intramural hematoma was often observed in a careful review in our corelab of 67 CTO procedures from 4 Japanese centers, suggesting that the guidewire frequently entered the medial space during successful recanalization.[125] CTO length as measured with angiography was shorter than the IVUS length. In a report of 219 patients with successfully recanalized CTO followed by IVUS, subintimal tracking was detected in 52% of cases, more often when using dissection reentry than wire escalation (87% vs 28%, respectively). In the subintimal tracking group, there was a higher rate of MACE, mostly driven by periprocedural myocardial infarction and there was a significantly greater incidence of angiographic dye staining/extravasation, and branch occlusion.[126] After stenting, struts expansion and apposition can be optimized with IVUS assessment, improving clinical outcomes. Among 201 propensity-score matched pairs included in the Korean-CTO Registry, IVUS-guided PCI was correlated with a lower trend of myocardial infarction and less stent thrombosis compared to angiography-guided procedure during a 2-year follow-up.[127] In the first randomized study comparing IVUS-guided with conventional angiography-guided CTO intervention using new-generation DES, 402 patients after wire crossing of the CTOs were randomized to the IVUS-guided group or the angiography-guided group.[128] At 12-month follow-up, there was a lower rate of MACE rates in the IVUS-guided group

(2.6% vs 7.1%; $P=.035$; HR: 0.35; 95% CI: 0.13–0.97). The AIR-CTO study also found that the rate of definite and/or probable stent thrombosis at 2-year follow-up was significantly lower among patients with IVUS guidance.[129] As often happens in interventional cardiology, subsequent to well-conducted randomized trials demonstrating a long-term beneficial impact of IVUS guidance, the multicentric PROGRESS CTO registry did not find, in 619 CTO PCIs where IVUS was used in 38% of the 606 patients, a difference in crossing or in procedural success, nor in-hospital MACE,[130] so many interventionalists are not convinced that IVUS guidance is useful in this subset of patients, while a position paper from experts recommends its use.[8]

As illustrated in Fig. 5, IVUS can also be used to resolve proximal cap ambiguity and guide the puncture of stumpless lesions by imaging through a side branch adjacent to the occlusion. In general, after lesion crossing, IVUS will confirm the presence of the wire in the distal true lumen. It might also be helpful to navigate the guidewire back in the true lumen in case of a dissection and subintimal tracking, using the IVUS catheter in the subintimal space to guide antegrade reentry in the true lumen. This highly complex method is only recommended as a last step in the latest algorithm proposed to treat CTO,[131] recommending careful selection for a retrograde strategy in which IVUS is also very useful. The preferred retrograde recanalization strategy is based on the (reverse) controlled antegrade and retrograde subintimal tracking (CART) technique for wire reentry in the true proximal lumen. IVUS can show (i) the respective position of the antegrade and retrograde wire intraplaque or in the subintimal space, (ii) the size of the vessel for optimal balloon selection for the CART, and (iii) the best location for making the connection of the wires.[132,133]

However, the use of IVUS in CTO PCI has several limitations. Firstly, the presence of calcium obscures the position of the occlusion stump or impairs the detection of the collapsed true lumen. Furthermore, the probe is side-looking, so that the catheter must be inserted into the occlusion to image it. New catheters are under development with forward-looking design that might offer new recanalization strategies.[134,135]

### 7. IVUS PITFALLS

Despite the profound advantages of IVUS in the assessment of atherosclerosis in vivo, the major limitation is mainly related to the fact that it is invasive. In order to provide its unique information, it is mandatory to be held in a catheterization laboratory under experienced and well-trained operators and staff. Prolonged radiation exposure and increased contrast usage should

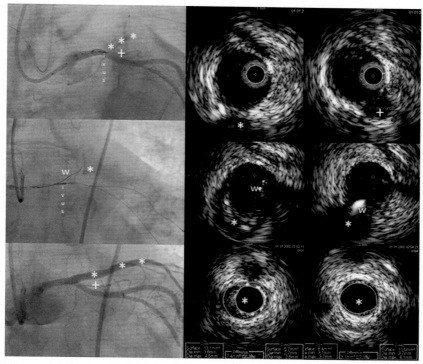

**FIG. 5** IVUS performed during a CTO PCI procedure. A totally occluded left anterior descending (LAD) artery at the level of the ostium is pointed on the angiogram on the *top left* by \*\*\*. *Top right*, on the corresponding IVUS cross section obtained just distally to the origin of the LAD, with the catheter in the left circumflex artery (Cx), a small side branch is marked by +, also visible on the angiogram. The IVUS cross section in the middle just more proximally shows the origin of the LAD (\*). *Middle panel, left*, shows the angiogram during the function of the proximal cap of the LAD with a wire (w) that can also be followed on the IVUS images in real time. The *lower panel* shows the stented LAD and the IVUS assessment of stent expansion in the LAD (*middle panel*) and its distal reference.

be also taken into consideration. From a technical point of view, the need to catheterize each vessel individually is also a matter of time and concern and relies always on the experience and skills of the interventional cardiologist. Anatomically speaking, another restriction is related to limited capabilities of imaging small-diameter vessels and aorto-ostial lesions.[61]

As with any visualization modality, certain artifacts may occur such as ring down, geometric distortion effect, blood speckle, nonuniform rotational distortion, or even broken catheters and devices. The drawback of high-frequency IVUS systems is that the intensity of the blood speckle increases to the fourth power of the transducer frequency so that the echodensity of blood might become as high as the plaque, if not higher in the presence of blood stasis with rouleau red-cell formation. Electronic filtering and image processing might reduce this phenomenon.[136] Moreover, intravascular

imaging sequences recorded in vivo suffer from motion artifacts mainly related to the beating heart. Attempts have been made to acquire electrocardiogram (ECG) gated frames with dedicated pullback devices. Frames are collected at the points in time associated with a particular fraction of the cardiac cycle. However, there are difficulties to determine the optimal fraction at which to gate, subsequent to the rather bulky specific pullback device to use. To circumvent this problem, frame-gating methods for IVUS pullbacks that mimic ECG gating (i.e., that select only one frame per cardiac cycle) have been developed. The algorithm automatically selects the fraction of the cycle that renders the most stable gated frame set.[137] Finally, calcifications produce a strong reflection of the ultrasound beam hampering the evaluation of deeper atherosclerosis.[68]

Another major concern is that image analysis should always be performed by well-trained experts with thorough understanding of this imaging field, otherwise it might lead to inaccurate measurements and mislead interpretation. The high cost and the occasionally limited availability due to approval or distribution issues remain a restriction to its worldwide spread.[61]

## 8. IVUS COMPLICATIONS

IVUS should be preceded by the administration of anticoagulants and intravenous nitrates in order to avoid vasospasm. Clinical studies have shown that IVUS can be performed safely with a low incidence of side effects (mainly local vasospasm), of the order of 1%. In a multicenter report of more than 2000 patients, common complications were vasospasm, acute occlusion, embolism, dissection, and thrombus formation with some patients presenting major events such as myocardial infarction or emergency coronary artery bypass surgery. The complication rate was higher in case of ACS (2.1%) as compared with patients with stable angina pectoris (0.8%) and asymptomatic patients (0.4%).[138,139] In the PROSPECT trial, 1.6% of events (10 coronary dissections and 1 perforation) were recorded among 697 ACS patients.[64] In 103 STEMI subjects who underwent three-vessel coronary imaging during PCI for the IBIS-4 study,[140] imaging of the noninfarct-related vessels was successful in approximately 90% of them without impact on cardiovascular events at long-term follow-up. Another recent report in about 2500 IVUS procedures was even more reassuring, with an IVUS-related complications rate of 0.5%. Complications were self-limiting after the retrieval of the imaging catheter or easily treatable in the catheterization laboratory without MACE.[141]

## 9. FUTURE PERSPECTIVES

Available data underline the evidence that IVUS is an excellent diagnostic imaging tool and with the latest DES multiple observational and randomized clinical trials demonstrated the potential of IVUS guidance to improve PCI outcomes. IVUS guidance for PCI remains recommended only in "selected" patients, with a class IIa recommendation for assessing the severity of a LMCA stenosis or for understanding stent-related mechanical problems leading to restenosis. It has also a class IIa in selected patients to optimize stent implantation, specifically left main lesions.[10] It is for sure an additional burden and it increases procedural time and might be difficult to implement for all procedures in a busy catheterization laboratory without any reimbursement, as quoted by Carlo Di Mario in his editorial about the ULTIMATE trial, "in view of the totality of RCT evidence,

there is no question that the use of IVUS guidance to optimize PCI does improve patient prognosis. Against this background, there is no scientific justification for the observed inertia in integrating an imaging-guided strategy more broadly in clinical practice."[25]

For the future, hybrid intravascular imaging based on the use of both OCT and IVUS has been recently proposed to better characterize lesion components and provide a detailed evaluation of the atherosclerotic plaque biology. This technique using multimodal catheters is expected to constitute an interesting approach for the study of atherosclerosis.[142] Additional efforts may include the development of a magnetic resonance catheter-based system that can identify lipid-rich tissue or even imaging catheters able to measure thermal gradients associated with inflammation in the coronary arteries. An ideal futuristic concept involves the potential use of a single catheter and pullback with the fusion of NIRS-IVUS in 3D. Finally, molecular imaging agents may enhance the identification of specific molecular processes within the plaques.

## DISCLOSURES

SC is a consultant for Boston Scientific and Terumo.

## REFERENCES

1. Topol EJ, Nissen SE. Our preoccupation with coronary luminology. The dissociation between clinical and angiographic findings in ischemic heart disease. *Circulation.* 1995; 92:2333–2342.
2. Yock PG, Fitzgerald PJ, Linker DT, Angelsen BA. Intravascular ultrasound guidance for catheter-based coronary interventions. *J Am Coll Cardiol.* 1991; 17:39B–45B.
3. Bom N, Lancee CT, Van Egmond FC. An ultrasonic intracardiac scanner. *Ultrasonics.* 1972; 10:72–76.
4. Edler I, Hertz CH. The use of ultrasonic reflectoscope for the continuous recording of the movements of heart walls. 1954. *Clin Physiol Funct Imaging.* 2004; 24:118–136.
5. Sanidas EA, Vavuranakis M, Papaioannou TG, et al. Study of atheromatous plaque using intravascular ultrasound. *Hell J Cardiol.* 2008; 49:415–421.
6. Johnson TW, Raber L, di Mario C, et al. Clinical use of intracoronary imaging. Part 2: acute coronary syndromes, ambiguous coronary angiography findings, and guiding interventional decision-making: an expert consensus document of the European Association of Percutaneous Cardiovascular Interventions. *Eur Heart J.* 2019; 40:2566–2584.
7. Song HG, Kang SJ, Mintz GS. Value of intravascular ultrasound in guiding coronary interventions. *Echocardiography.* 2018; 35:520–533.

8. Raber L, Mintz GS, Koskinas KC, et al. Clinical use of intracoronary imaging. Part 1: guidance and optimization of coronary interventions. An expert consensus document of the European Association of Percutaneous Cardiovascular Interventions. *EuroIntervention*. 2018; 14:656–677.

9. Fujii K, Carlier SG, Mintz GS, et al. Stent underexpansion and residual reference segment stenosis are related to stent thrombosis after sirolimus-eluting stent implantation: an intravascular ultrasound study. *J Am Coll Cardiol*. 2005; 45:995–998.

10. Neumann FJ, Sousa-Uva M, Ahlsson A, et al. 2018 ESC/EACTS Guidelines on myocardial revascularization. *EuroIntervention*. 2019. 14:1435–1534

11. Mintz GS, Guagliumi G. Intravascular imaging in coronary artery disease. *Lancet*. 2017; 390:793–809.

12. Bom N, Carlier SG, van der Steen AF, Lancee CT. Intravascular scanners. *Ultrasound Med Biol*. 2000; 26 (Suppl 1):S6–S9.

13. Velican D, Velican C. Comparative study on age-related changes and atherosclerotic involvement of the coronary arteries of male and female subjects up to 40 years of age. *Atherosclerosis*. 1981; 38:39–50.

14. Parviz Y, Shlofmitz E, Fall KN, et al. Utility of intracoronary imaging in the cardiac catheterization laboratory: comprehensive evaluation with intravascular ultrasound and optical coherence tomography. *Br Med Bull*. 2018; 125:79–90.

15. Mintz GS, Nissen SE, Anderson WD, et al. American College of Cardiology clinical expert consensus document on standards for acquisition, measurement and reporting of intravascular ultrasound studies (IVUS). A report of the American College of Cardiology Task Force on Clinical Expert Consensus Documents. *J Am Coll Cardiol*. 2001; 37:1478–1492.

16. Ross R. Atherosclerosis—an inflammatory disease. *N Engl J Med*. 1999; 340:115–126.

17. Glagov S, Weisenberg E, Zarins CK, Stankunavicius R, Kolettis GJ. Compensatory enlargement of human atherosclerotic coronary arteries. *N Engl J Med*. 1987; 316:1371–1375.

18. Ge J, Erbel R, Gerber T, et al. Intravascular ultrasound imaging of angiographically normal coronary arteries: a prospective study in vivo. *Br Heart J*. 1994; 71:572–578.

19. Ambrose JA, Tannenbaum MA, Alexopoulos D, et al. Angiographic progression of coronary artery disease and the development of myocardial infarction. *J Am Coll Cardiol*. 1988; 12:56–62.

20. Schroeder AP, Falk E. Vulnerable and dangerous coronary plaques. *Atherosclerosis*. 1995; 118(Suppl):S141–S149.

21. Nair A, Kuban BD, Tuzcu EM, Schoenhagen P, Nissen SE, Vince DG. Coronary plaque classification with intravascular ultrasound radiofrequency data analysis. *Circulation*. 2002; 106:2200–2206.

22. Okubo M, Kawasaki M, Ishihara Y, et al. Development of integrated backscatter intravascular ultrasound for tissue characterization of coronary plaques. *Ultrasound Med Biol*. 2008; 34:655–663.

23. Sathyanarayana S, Carlier S, Li W, Thomas L. Characterisation of atherosclerotic plaque by spectral similarity of radiofrequency intravascular ultrasound signals. *EuroIntervention*. 2009; 5:133–139.

24. Katouzian A, Karamalis A, Sheet D, et al. Iterative self-organizing atherosclerotic tissue labeling in intravascular ultrasound images and comparison with virtual histology. *IEEE Trans Biomed Eng*. 2012; 59:3039–3049.

25. di Mario C, Koskinas KC, Raber L. Clinical benefit of IVUS guidance for coronary stenting: the ULTIMATE step toward definitive evidence? *J Am Coll Cardiol*. 2018; 72:3138–3141.

26. Stone GW, Hodgson JM, St Goar FG, et al. Improved procedural results of coronary angioplasty with intravascular ultrasound-guided balloon sizing: the CLOUT Pilot Trial. Clinical Outcomes With Ultrasound Trial (CLOUT) Investigators. *Circulation*. 1997; 95: 2044–2052.

27. Haase KK, Athanasiadis A, Mahrholdt H, et al. Acute and one year follow-up results after vessel size adapted PTCA using intracoronary ultrasound. *Eur Heart J*. 1998; 19:263–272.

28. Kimura T, Kaburagi S, Tamura T, et al. Remodeling of human coronary arteries undergoing coronary angioplasty or atherectomy. *Circulation*. 1997; 96:475–483.

29. Serruys PW, de Bruyne B, Carlier S, et al. Randomized comparison of primary stenting and provisional balloon angioplasty guided by flow velocity measurement. Doppler Endpoints Balloon Angioplasty Trial Europe (DEBATE) II Study Group. *Circulation*. 2000; 102:2930–2937.

30. de Ribamar Costa Jr. J, Mintz GS, Carlier SG, et al. Nonrandomized comparison of coronary stenting under intravascular ultrasound guidance of direct stenting without predilation versus conventional predilation with a semi-compliant balloon versus predilation with a new scoring balloon. *Am J Cardiol*. 2007; 100:812–817.

31. Barbato E, Carrie D, Dardas P, et al. European expert consensus on rotational atherectomy. *EuroIntervention*. 2015; 11:30–36.

32. Sanidas EAMG, Maehara A, Choi S, et al. Intracoronary ultrasound for optimizing stent implantation. *Curr Cardiovasc Imaging Rep*. 2010; 3:230–236.

33. Fitzgerald PJ, Oshima A, Hayase M, et al. Final results of the Can Routine Ultrasound Influence Stent Expansion (CRUISE) study. *Circulation*. 2000; 102:523–530.

34. Oemrawsingh PV, Mintz GS, Schalij MJ, et al. Intravascular ultrasound guidance improves angiographic and clinical outcome of stent implantation for long coronary artery stenoses: final results of a randomized comparison with angiographic guidance (TULIP Study). *Circulation*. 2003; 107:62–67.

35. Gil RJ, Pawlowski T, Dudek D, et al. Comparison of angiographically guided direct stenting technique with direct stenting and optimal balloon angioplasty guided

with intravascular ultrasound. The multicenter, randomized trial results. *Am Heart J.* 2007; 154:669–675.

36. Russo RJ, Silva PD, Teirstein PS, et al. A randomized controlled trial of angiography versus intravascular ultrasound-directed bare-metal coronary stent placement (the AVID Trial). *Circ Cardiovasc Interv.* 2009; 2:113–123.

37. de Jaegere P, Mudra H, Figulla H, et al. Intravascular ultrasound-guided optimized stent deployment. Immediate and 6 months clinical and angiographic results from the Multicenter Ultrasound Stenting in Coronaries Study (MUSIC Study). *Eur Heart J.* 1998; 19:1214–1223.

38. Parise H, Maehara A, Stone GW, Leon MB, Mintz GS. Meta-analysis of randomized studies comparing intravascular ultrasound versus angiographic guidance of percutaneous coronary intervention in pre-drug-eluting stent era. *Am J Cardiol.* 2011; 107:374–382.

39. Mudra H, di Mario C, de Jaegere P, et al. Randomized comparison of coronary stent implantation under ultrasound or angiographic guidance to reduce stent restenosis (OPTICUS Study). *Circulation.* 2001; 104:1343–1349.

40. Casella G, Klauss V, Ottani F, Siebert U, Sangiorgio P, Bracchetti D. Impact of intravascular ultrasound-guided stenting on long-term clinical outcome: a meta-analysis of available studies comparing intravascular ultrasound-guided and angiographically guided stenting. *Catheter Cardiovasc Interv.* 2003; 59:314–321.

41. Choi SY, Maehara A, Cristea E, et al. Usefulness of minimum stent cross sectional area as a predictor of angiographic restenosis after primary percutaneous coronary intervention in acute myocardial infarction (from the HORIZONS-AMI trial IVUS substudy). *Am J Cardiol.* 2012; 109:455–460.

42. Choi SY, Witzenbichler B, Maehara A, et al. Intravascular ultrasound findings of early stent thrombosis after primary percutaneous intervention in acute myocardial infarction: a Harmonizing Outcomes with Revascularization and Stents in Acute Myocardial Infarction (HORIZONS-AMI) substudy. *Circ Cardiovasc Interv.* 2011; 4:239–247.

43. Claessen BE, Mehran R, Mintz GS, et al. Impact of intravascular ultrasound imaging on early and late clinical outcomes following percutaneous coronary intervention with drug-eluting stents. *JACC Cardiovasc Interv.* 2011; 4:974–981.

44. Dangas GD, Claessen BE, Caixeta A, Sanidas EA, Mintz GS, Mehran R. In-stent restenosis in the drug-eluting stent era. *J Am Coll Cardiol.* 2010; 56:1897–1907.

45. Takebayashi H, Mintz GS, Carlier SG, et al. Nonuniform strut distribution correlates with more neointimal hyperplasia after sirolimus-eluting stent implantation. *Circulation.* 2004; 110:3430–3434.

46. Steinberg DH, Mintz GS, Mandinov L, et al. Long-term impact of routinely detected early and late incomplete stent apposition: an integrated intravascular ultrasound

analysis of the TAXUS IV, V, and VI and TAXUS ATLAS workhorse, long lesion, and direct stent studies. *JACC Cardiovasc Interv.* 2010; 3:486–494.

47. Zhang Y, Farooq V, Garcia-Garcia HM, et al. Comparison of intravascular ultrasound versus angiography-guided drug-eluting stent implantation: a meta-analysis of one randomised trial and ten observational studies involving 19,619 patients. *EuroIntervention.* 2012; 8:855–865.

48. Hong SJ, Kim BK, Shin DH, et al. Effect of intravascular ultrasound-guided vs angiography-guided everolimus-eluting stent implantation: the IVUS-XPL randomized clinical trial. *JAMA.* 2015; 314:2155–2163.

49. Steinvil A, Zhang YJ, Lee SY, et al. Intravascular ultrasound-guided drug-eluting stent implantation: an updated meta-analysis of randomized control trials and observational studies. *Int J Cardiol.* 2016; 216:133–139.

50. Elgendy IY, Mahmoud AN, Elgendy AY, Bavry AA. Outcomes with intravascular ultrasound-guided stent implantation: a meta-analysis of randomized trials in the era of drug-eluting stents. *Circ Cardiovasc Interv.* 2016; 9. e003700.

51. Witzenbichler B, Maehara A, Weisz G, et al. Relationship between intravascular ultrasound guidance and clinical outcomes after drug-eluting stents: the assessment of dual antiplatelet therapy with drug-eluting stents (ADAPT-DES) study. *Circulation.* 2014; 129:463–470.

52. Maehara A, Mintz GS, Witzenbichler B, et al. Relationship between intravascular ultrasound guidance and clinical outcomes after drug-eluting stents. *Circ Cardiovasc Interv.* 2018; 11. e006243.

53. Zhang J, Gao X, Kan J, et al. Intravascular ultrasound versus angiography-guided drug-eluting stent implantation: the ULTIMATE trial. *J Am Coll Cardiol.* 2018; 72:3126–3137.

54. Katagiri Y, De Maria GL, Kogame N, et al. Impact of post-procedural minimal stent area on 2-year clinical outcomes in the SYNTAX II trial. *Catheter Cardiovasc Interv.* 2019; 93:E225–E234.

55. Choi KH, Song YB, Lee JM, et al. Impact of intravascular ultrasound-guided percutaneous coronary intervention on long-term clinical outcomes in patients undergoing complex procedures. *JACC Cardiovasc Interv.* 2019; 12 (7):607–620.

56. Toutouzas K, Stefanadis C. Advances in vulnerable plaque detection and treatment: how far have we gone? *Hell J Cardiol.* 2006; 47:129–131.

57. Schwartz SM, Virmani R, Rosenfeld ME. The good smooth muscle cells in atherosclerosis. *Curr Atheroscler Rep.* 2000; 2:422–429.

58. Kawasaki M, Takatsu H, Noda T, et al. Noninvasive quantitative tissue characterization and two-dimensional color-coded map of human atherosclerotic lesions using ultrasound integrated backscatter: comparison between histology and integrated backscatter images. *J Am Coll Cardiol.* 2001; 38:486–492.

59. Kawasaki M, Takatsu H, Noda T, et al. In vivo quantitative tissue characterization of human coronary arterial

plaques by use of integrated backscatter intravascular ultrasound and comparison with angioscopic findings. *Circulation.* 2002; 105:2487–2492.

60. Ohota M, Kawasaki M, Ismail TF, Hattori K, Serruys PW, Ozaki Y. A histological and clinical comparison of new and conventional integrated backscatter intravascular ultrasound (IB-IVUS). *Circ J.* 2012; 76: 1678–1686.

61. Sanidas E, Dangas G. Evolution of intravascular assessment of coronary anatomy and physiology: from ultrasound imaging to optical and flow assessment. *Eur J Clin Investig.* 2013; 43:996–1008.

62. Sanidas EA, Maehara A, Mintz GS, et al. Angioscopic and virtual histology intravascular ultrasound characteristics of culprit lesion morphology underlying coronary artery thrombosis. *Am J Cardiol.* 2011; 107:1285–1290.

63. Thim T, Hagensen MK, Wallace-Bradley D, et al. Unreliable assessment of necrotic core by virtual histology intravascular ultrasound in porcine coronary artery disease. *Circ Cardiovasc Imaging.* 2010; 3:384–391.

64. Stone GW, Maehara A, Lansky AJ, et al. A prospective natural-history study of coronary atherosclerosis. *N Engl J Med.* 2011; 364:226–235.

65. Calvert PA, Obaid DR, O'Sullivan M, et al. Association between IVUS findings and adverse outcomes in patients with coronary artery disease: the VIVA (VH-IVUS in Vulnerable Atherosclerosis) Study. *JACC Cardiovasc Imaging.* 2011; 4:894–901.

66. Seo YH, Kim YK, Song IG, Kim KH, Kwon TG, Bae JH. Long-term clinical outcomes in patients with untreated non-culprit intermediate coronary lesion and evaluation of predictors by using virtual histology-intravascular ultrasound; a prospective cohort study. *BMC Cardiovasc Disord.* 2019; 19:187.

67. Katouzian A, Sathyanarayana S, Baseri B, Konofagou EE, Carlier SG. Challenges in atherosclerotic plaque characterization with intravascular ultrasound (IVUS): from data collection to classification. *IEEE Trans Inf Technol Biomed.* 2008; 12:315–327.

68. Sheet D, Karamalis A, Eslami A, et al. Hunting for necrosis in the shadows of intravascular ultrasound. *Comput Med Imaging Graph.* 2014; 38:104–112.

69. Erglis A, Jegere S, Narbute I. Intravascular ultrasound-based imaging modalities for tissue characterisation. *Interv Cardiol.* 2014; 9:151–155.

70. Garcia-Garcia HM, Gogas BD, Serruys PW, Bruining N. IVUS-based imaging modalities for tissue characterization: similarities and differences. *Int J Card Imaging.* 2011; 27:215–224.

71. Kozuki A, Shinke T, Otake H, et al. Feasibility of a novel radiofrequency signal analysis for in-vivo plaque characterization in humans: comparison of plaque components between patients with and without acute coronary syndrome. *Int J Cardiol.* 2013; 167:1591–1596.

72. Trusinskis K, Juhnevica D, Strenge K, Erglis A. iMap intravascular ultrasound evaluation of culprit and non-culprit lesions in patients with ST-elevation myocardial infarction. *Cardiovasc Revasc Med.* 2013; 14:71–75.

73. Utsunomiya M, Hara H, Sugi K, Nakamura M. Relationship between tissue characterisations with 40 MHz intravascular ultrasound imaging and slow flow during coronary intervention. *EuroIntervention.* 2011; 7:340–346.

74. Tsujita K, Takaoka N, Kaikita K, et al. Neointimal tissue component assessed by tissue characterization with 40 MHz intravascular ultrasound imaging: comparison of drug-eluting stents and bare-metal stents. *Catheter Cardiovasc Interv.* 2013; 82:1068–1074.

75. Gardner CM, Tan H, Hull EL, et al. Detection of lipid core coronary plaques in autopsy specimens with a novel catheter-based near-infrared spectroscopy system. *JACC Cardiovasc Imaging.* 2008; 1:638–648.

76. Goldstein JA, Maini B, Dixon SR, et al. Detection of lipid-core plaques by intracoronary near-infrared spectroscopy identifies high risk of periprocedural myocardial infarction. *Circ Cardiovasc Interv.* 2011; 4:429–437.

77. Stone GW, Maehara A, Muller JE, et al. Plaque characterization to inform the prediction and prevention of periprocedural myocardial infarction during percutaneous coronary intervention: the CANARY trial (coronary assessment by near-infrared of atherosclerotic rupture-prone yellow). *JACC Cardiovasc Interv.* 2015; 8:927–936.

78. Waksman R, Di Mario C, Torguson R, et al. Identification of patients and plaques vulnerable to future coronary events with near-infrared spectroscopy intravascular ultrasound imaging: a prospective, cohort study. *Lancet.* 2019; 394:1629–1637.

79. Carlier S, Kakadiaris IA, Dib N, et al. Vasa vasorum imaging: a new window to the clinical detection of vulnerable atherosclerotic plaques. *Curr Atheroscler Rep.* 2005; 7:164–169.

80. Vavuranakis M, Kakadiaris IA, Papaioannou TG, et al. Contrast-enhanced intravascular ultrasound: combining morphology with activity-based assessment of plaque vulnerability. *Expert Rev Cardiovasc Ther.* 2007; 5:917–925.

81. Vavuranakis M, Kakadiaris IA, O'Malley SM, et al. A new method for assessment of plaque vulnerability based on vasa vasorum imaging, by using contrast-enhanced intravascular ultrasound and differential image analysis. *Int J Cardiol.* 2008; 130:23–29.

82. Ruiz EM, Papaioannou TG, Vavuranakis M, Stefanadis C, Naghavi M, Kakadiaris IA. Analysis of contrast-enhanced intravascular ultrasound images for the assessment of coronary plaque neoangiogenesis: another step closer to the identification of the vulnerable plaque. *Curr Pharm Des.* 2012; 18:2207–2213.

83. Vavuranakis M, Papaioannou TG, Kakadiaris IA, et al. Detection of perivascular blood flow in vivo by contrast-enhanced intracoronary ultrasonography and image analysis: an animal study. *Clin Exp Pharmacol Physiol.* 2007; 34:1319–1323.

84. Vavuranakis M, Papaioannou TG, Vrachatis D, et al. Computational imaging of aortic vasa vasorum and neovascularization in rabbits using contrast-enhanced

intravascular ultrasound: association with histological analysis. *Anatol J Cardiol.* 2018; 20:117–124.

85. Kawasaki M, Sano K, Okubo M, et al. Volumetric quantitative analysis of tissue characteristics of coronary plaques after statin therapy using three-dimensional integrated backscatter intravascular ultrasound. *J Am Coll Cardiol.* 2005; 45:1946–1953.

86. Nissen SE, Tuzcu EM, Schoenhagen P, et al. Effect of intensive compared with moderate lipid-lowering therapy on progression of coronary atherosclerosis: a randomized controlled trial. *JAMA.* 2004; 291:1071–1080.

87. Okazaki S, Yokoyama T, Miyauchi K, et al. Early statin treatment in patients with acute coronary syndrome: demonstration of the beneficial effect on atherosclerotic lesions by serial volumetric intravascular ultrasound analysis during half a year after coronary event: the ESTABLISH Study. *Circulation.* 2004; 110:1061–1068.

88. Nissen SE, Nicholls SJ, Sipahi I, et al. Effect of very high-intensity statin therapy on regression of coronary atherosclerosis: the ASTEROID trial. *JAMA.* 2006; 295:1556–1565.

89. Tsujita K, Sugiyama S, Sumida H, et al. Impact of dual lipid-lowering strategy with ezetimibe and atorvastatin on coronary plaque regression in patients with percutaneous coronary intervention: the multicenter randomized controlled PRECISE-IVUS trial. *J Am Coll Cardiol.* 2015; 66:495–507.

90. Nicholls SJ, Puri R, Anderson T, et al. Effect of evolocumab on progression of coronary disease in statin-treated patients: the GLAGOV randomized clinical trial. *JAMA.* 2016; 316:2373–2384.

91. Nicholls SJ, Puri R, Anderson T, et al. Effect of evolocumab on coronary plaque composition. *J Am Coll Cardiol.* 2018; 72:2012–2021.

92. Ako J, Hibi K, Kozuma K, et al. Effect of alirocumab on coronary atheroma volume in Japanese patients with acute coronary syndromes and hypercholesterolemia not adequately controlled with statins: ODYSSEY J-IVUS rationale and design. *J Cardiol.* 2018; 71:583–589.

93. Chieffo A, Latib A, Caussin C, et al. A prospective, randomized trial of intravascular-ultrasound guided compared to angiography guided stent implantation in complex coronary lesions: the AVIO trial. *Am Heart J.* 2013; 165:65–72.

94. Bavishi C, Sardar P, Chatterjee S, et al. Intravascular ultrasound-guided vs angiography-guided drug-eluting stent implantation in complex coronary lesions: meta-analysis of randomized trials. *Am Heart J.* 2017; 185:26–34.

95. Shin DH, Hong SJ, Mintz GS, et al. Effects of intravascular ultrasound-guided versus angiography-guided new-generation drug-eluting stent implantation: meta-analysis with individual patient-level data from 2,345 randomized patients. *JACC Cardiovasc Interv.* 2016; 9:2232–2239.

96. Sano K, Mintz GS, Carlier SG, et al. Assessing intermediate left main coronary lesions using intravascular ultrasound. *Am Heart J.* 2007; 154:983–988.

97. Fassa AA, Wagatsuma K, Higano ST, et al. Intravascular ultrasound-guided treatment for angiographically indeterminate left main coronary artery disease: a long-term follow-up study. *J Am Coll Cardiol.* 2005; 45:204–211.

98. Jasti V, Ivan E, Yalamanchili V, Wongpraparut N, Leesar MA. Correlations between fractional flow reserve and intravascular ultrasound in patients with an ambiguous left main coronary artery stenosis. *Circulation.* 2004; 110:2831–2836.

99. Kang SJ, Lee JY, Ahn JM, et al. Intravascular ultrasound-derived predictors for fractional flow reserve in intermediate left main disease. *JACC Cardiovasc Interv.* 2011; 4:1168–1174.

100. Park SJ, Ahn JM, Kang SJ, et al. Intravascular ultrasound-derived minimal lumen area criteria for functionally significant left main coronary artery stenosis. *JACC Cardiovasc Interv.* 2014; 7:868–874.

101. Rusinova RP, Mintz GS, Choi SY, et al. Intravascular ultrasound comparison of left main coronary artery disease between white and Asian patients. *Am J Cardiol.* 2013; 111:979–984.

102. Mintz GS, Lefevre T, Lassen JF, et al. Intravascular ultrasound in the evaluation and treatment of left main coronary artery disease: a consensus statement from the European Bifurcation Club. *EuroIntervention.* 2018; 14: e467–e474.

103. Hamilos M, Muller O, Cuisset T, et al. Long-term clinical outcome after fractional flow reserve-guided treatment in patients with angiographically equivocal left main coronary artery stenosis. *Circulation.* 2009; 120:1505–1512.

104. de la Torre Hernandez JM, Hernandez Hernandez F, Alfonso F, et al. Prospective application of pre-defined intravascular ultrasound criteria for assessment of intermediate left main coronary artery lesions results from the multicenter LITRO study. *J Am Coll Cardiol.* 2011; 58:351–358.

105. Park SJ, Kim YH, Park DW, et al. Impact of intravascular ultrasound guidance on long-term mortality in stenting for unprotected left main coronary artery stenosis. *Circ Cardiovasc Interv.* 2009; 2:167–177.

106. Maehara A, Mintz G, Serruys P, et al. Impact of final minimal stent area by IVUS on 3-year outcome after PCI of left main coronary artery disease: the EXCEL trial. *J Am Coll Cardiol.* 2017; 69(11 Suppl). https://doi.org/10.1016/S0735-1097(17)34352-8.

107. Makikallio T, Holm NR, Lindsay M, et al. Percutaneous coronary angioplasty versus coronary artery bypass grafting in treatment of unprotected left main stenosis (NOBLE): a prospective, randomised, open-label, non-inferiority trial. *Lancet.* 2016; 388:2743–2752.

108. Fajadet J, Capodanno D, Stone GW. Management of left main disease: an update. *Eur Heart J.* 2019. 40 (18):1454–1466

109. Abizaid A, Mintz GS, Pichard AD, et al. Clinical, intravascular ultrasound, and quantitative angiographic determinants of the coronary flow reserve before and after percutaneous transluminal coronary angioplasty. *Am J Cardiol.* 1998; 82:423–428.

110. Nishioka T, Amanullah AM, Luo H, et al. Clinical validation of intravascular ultrasound imaging for assessment of coronary stenosis severity: comparison with stress myocardial perfusion imaging. *J Am Coll Cardiol.* 1999; 33:1870–1878.

111. Mintz GS. Clinical utility of intravascular imaging and physiology in coronary artery disease. *J Am Coll Cardiol.* 2014; 64:207–222.

112. Abizaid AS, Mintz GS, Mehran R, et al. Long-term follow-up after percutaneous transluminal coronary angioplasty was not performed based on intravascular ultrasound findings: importance of lumen dimensions. *Circulation.* 1999; 100:256–261.

113. Nam CW, Yoon HJ, Cho YK, et al. Outcomes of percutaneous coronary intervention in intermediate coronary artery disease: fractional flow reserve-guided versus intravascular ultrasound-guided. *JACC Cardiovasc Interv.* 2010; 3:812–817.

114. Kang SJ, Ahn JM, Song H, et al. Usefulness of minimal luminal coronary area determined by intravascular ultrasound to predict functional significance in stable and unstable angina pectoris. *Am J Cardiol.* 2012; 109:947–953.

115. Kang J, Koo BK, Hu X, et al. Comparison of Fractional FLow Reserve And Intravascular ultrasound-guided Intervention Strategy for Clinical OUtcomes in Patients with InteRmediate Stenosis (FLAVOUR): rationale and design of a randomized clinical trial. *Am Heart J.* 2018; 199:7–12.

116. Mintz GS, Popma JJ, Pichard AD, et al. Patterns of calcification in coronary artery disease. A statistical analysis of intravascular ultrasound and coronary angiography in 1155 lesions. *Circulation.* 1995; 91:1959–1965.

117. Mintz GS. Intravascular imaging of coronary calcification and its clinical implications. *JACC Cardiovasc Imaging.* 2015; 8:461–471.

118. Fujii K, Carlier SG, Mintz GS, et al. Intravascular ultrasound study of patterns of calcium in ruptured coronary plaques. *Am J Cardiol.* 2005; 96:352–357.

119. Vengrenyuk Y, Carlier S, Xanthos S, et al. A hypothesis for vulnerable plaque rupture due to stress-induced debonding around cellular microcalcifications in thin fibrous caps. *Proc Natl Acad Sci U S A.* 2006; 103:14678–14683.

120. Strauss HW, Nakahara T, Narula N, Narula J. Vascular calcification: the evolving relationship of vascular calcification to major acute coronary events. *J Nucl Med.* 2019; 60:1207–1212.

121. Galassi AR, Sumitsuji S, Boukhris M, et al. Utility of intravascular ultrasound in percutaneous

revascularization of chronic total occlusions: an overview. *JACC Cardiovasc Interv.* 2016; 9:1979–1991.

122. Werner GS, Diedrich J, Scholz KH, Knies A, Kreuzer H. Vessel reconstruction in total coronary occlusions with a long subintimal wire pathway: use of multiple stents under guidance of intravascular ultrasound. *Catheter Cardiovasc Diagn.* 1997; 40:46–51.

123. Ito S, Suzuki T, Ito T, et al. Novel technique using intravascular ultrasound-guided guidewire cross in coronary intervention for uncrossable chronic total occlusions. *Circ J.* 2004; 68:1088–1092.

124. Escaned J. Percutaneous treatment of chronic total coronary occlusions: the light that came from Japan. *JACC Cardiovasc Interv.* 2017; 10:2155–2157.

125. Fujii K, Ochiai M, Mintz GS, et al. Procedural implications of intravascular ultrasound morphologic features of chronic total coronary occlusions. *Am J Cardiol.* 2006; 97:1455–1462.

126. Song L, Maehara A, Finn MT, et al. Intravascular ultrasound analysis of intraplaque versus subintimal tracking in percutaneous intervention for coronary chronic total occlusions and association with procedural outcomes. *JACC Cardiovasc Interv.* 2017; 10:1011–1021.

127. Hong SJ, Kim BK, Shin DH, et al. Usefulness of intravascular ultrasound guidance in percutaneous coronary intervention with second-generation drug-eluting stents for chronic total occlusions (from the Multicenter Korean-Chronic Total Occlusion Registry). *Am J Cardiol.* 2014; 114:534–540.

128. Kim BK, Shin DH, Hong MK, et al. Clinical impact of intravascular ultrasound-guided chronic total occlusion intervention with zotarolimus-eluting versus biolimus-eluting stent implantation: randomized study. *Circ Cardiovasc Interv.* 2015; 8. e002592.

129. Tian NL, Gami SK, Ye F, et al. Angiographic and clinical comparisons of intravascular ultrasound- versus angiography-guided drug-eluting stent implantation for patients with chronic total occlusion lesions: two-year results from a randomised AIR-CTO study. *EuroIntervention.* 2015; 10:1409–1417.

130. Karacsonyi J, Alaswad K, Jaffer FA, et al. Use of intravascular imaging during chronic total occlusion percutaneous coronary intervention: insights from a contemporary multicenter registry. *J Am Heart Assoc.* 2016; 5(8). e003890. https://doi.org/10.1161/JAHA.116.003890.

131. Harding SA, Wu EB, Lo S, et al. A new algorithm for crossing chronic total occlusions from the Asia Pacific Chronic Total Occlusion Club. *JACC Cardiovasc Interv.* 2017; 10:2135–2143.

132. Rathore S, Katoh O, Tuschikane E, Oida A, Suzuki T, Takase S. A novel modification of the retrograde approach for the recanalization of chronic total occlusion of the coronary arteries intravascular ultrasound-guided reverse controlled antegrade and retrograde tracking. *JACC Cardiovasc Interv.* 2010; 3:155–164.

133. Matsuno S, Tsuchikane E, Harding SA, et al. Overview and proposed terminology for the reverse controlled antegrade and retrograde tracking (reverse CART) techniques. *EuroIntervention.* 2018; 14:94–101.

134. Gurun G, Tekes C, Zahorian J, et al. Single-chip CMUT-on-CMOS front-end system for real-time volumetric IVUS and ICE imaging. *IEEE Trans Ultrason Ferroelectr Freq Control.* 2014; 61:239–250.

135. Lee J, Chang JH. A 40-MHz ultrasound transducer with an angled aperture for guiding percutaneous revascularization of chronic total occlusion: a feasibility study. *Sensors (Basel).* 2018; 18(11):4079. https://doi.org/10.3390/s18114079.

136. Hibi K, Takagi A, Zhang X, et al. Feasibility of a novel blood noise reduction algorithm to enhance reproducibility of ultra-high-frequency intravascular ultrasound images. *Circulation.* 2000; 102:1657–1663.

137. O'Malley SM, Granada JF, Carlier S, Naghavi M, Kakadiaris IA. Image-based gating of intravascular ultrasound pullback sequences. *IEEE Trans Inf Technol Biomed.* 2008; 12:299–306.

138. Batkoff BW, Linker DT. Safety of intracoronary ultrasound: data from a Multicenter European Registry. *Catheter Cardiovasc Diagn.* 1996; 38:238–241.

139. Hausmann D, Erbel R, Alibelli-Chemarin MJ, et al. The safety of intracoronary ultrasound. A multicenter survey of 2207 examinations. *Circulation.* 1995; 91:623–630.

140. Taniwaki M, Radu MD, Garcia-Garcia HM, et al. Long-term safety and feasibility of three-vessel multimodality intravascular imaging in patients with ST-elevation myocardial infarction: the IBIS-4 (integrated biomarker and imaging study) substudy. *Int J Card Imaging.* 2015; 31:915–926.

141. van der Sijde JN, Karanasos A, van Ditzhuijzen NS, et al. Safety of optical coherence tomography in daily practice: a comparison with intravascular ultrasound. *Eur Heart J Cardiovasc Imaging.* 2017; 18:467–474.

142. Bourantas CV, Jaffer FA, Gijsen FJ, et al. Hybrid intravascular imaging: recent advances, technical considerations, and current applications in the study of plaque pathophysiology. *Eur Heart J.* 2017; 38:400–412.

# CHAPTER 3

# Convenience of Intravascular Ultrasound in Coronary Chronic Total Occlusion Recanalization

JUAN RIGLA[a,b] • JOSEP RIGLA[b] • FERNANDO RAMOS[b] • JOSEP LLUÍS GÓMEZ-HUERTAS[a] • LUKASZ PARTYKA[a] • MARIA ELENA DE CEGLIA[a] • JOAN ANTONI GOMEZ-HOSPITAL[c] • BEATRIZ VAQUERIZO[d]

[a]InspireMD, Tel-Aviv, Israel, [b]Department of Mathematics and Informatics, University of Barcelona, Barcelona, Spain, [c]Interventional Cardiology Unit, Hospital de Bellvitge, Barcelona, Spain, [d]Interventional Cardiology Unit, Hospital del Mar, Barcelona, Spain

Even today, chronic total occlusion (CTO) revascularization constitutes a procedural challenge. Tapered stump CTO lesions are the ones with better success share, while blunt and vague stump (stumpless) CTO lesions have worse achievement because of the difficulty of antero-grade to predict the guidewire course from the occluded point to the true distal lumen, especially those with a side branch arising from the occlusion. Still, when anterograde recanalization is not possible, retrograde recanalization constitutes an alternative challenge. Currently, the CTO lesions recanalization presents a high success rate (80%–90%) when reported by experienced operators.

Intravascular ultrasound (IVUS) imaging can assist in differentiating a true lumen against subintimal space by recognizing the presence of side branches, intima, and media (found in the true lumen, but not in the subintimal space). Besides, IVUS can confirm guidewire reentry to the true lumen from the subintimal space. IVUS-guided wiring technique may constitute one of the alternatives in the antegrade approach when standard wiring procedures fail.

In addition, anterograde IVUS may be beneficial in cases of retrograde approach.

The use of IVUS in CTO has a long history of pioneering investigators and interventional contributions. The recognition of evolvers will constitute Section 1 on this chapter. Section 2 will schematically summarize the IVUS convenience on CTO recanalization. Section 3 is dedicated to example cases, illustrative of the use of IVUS in CTO and their benefits.

## 1. INNOVATION CHRONOLOGY ON CTO CORONARY INTERVENTION GUIDED BY IVUS

The first to publish a serie on CTO treatment was Holmes in 1984, with a cohort of 26 patients.[1] One year later, Serruys was the first to state that elective percutaneous coronary intervention (PCI) is feasible in CTO, although with a low primary success rate (57% at that time). The long-term clinical follow-up also exposed a high incidence of reocclusion (40% with bare stents).[2]

A decade later appeared the first case-report of IVUS on CTO published by Kimura using a 3.5 F, 30 MHz Sonicath (BSC), operating a Hewlett Packard SONOS intravascular imaging system. This earliest case was a CTO angioplasty complicated with a subintimal wire dissection, where IVUS was used to provide online information on wire position to complement the angiographic image.[3] However, Matsubara in 2004 was the first to define a systematic IVUS approach for CTO. He reported two cases of IVUS-guided CTO wiring technique using stiffer wires and low-profile intracoronary ultrasound catheters. In the first case, a guidewire in subintimal space with the IVUS catheter was used to guide another wire into the true lumen. In the second case, an LAD occlusion wiring was enabled by placing the IVUS probe in the diagonal side branch. He was the first to state that IVUS may improve the success rate of CTO treatment.[4]

Importantly, Ozawa was the first to describe the CTO retrograde approach. The method consists in navigating

*Intravascular Ultrasound.* https://doi.org/10.1016/B978-0-12-818833-0.00003-5

the guidewire from the contralateral arterial bed to pass the CTO retrogradely and follow across the homolateral arterial bed to externalize the guidewire outside the body. He performed the first retrograde approach in two patients with right coronary artery (RCA) CTO via LAD septal branch.[5]

García-García published a review paper summarizing different techniques used in CTO recanalization. This concerns the use of IVUS to cross the occlusion or to reenter into the true lumen.[6] Cuneo issued a histopathological CTO description where he defined varying gradations of plaques, fibrous caps, and neovascularization. Cuneo also summarized the anterograde technical approach: "parallel" and "seesaw" wire techniques, balloon anchoring, subintimal tracking and reentry (STAR), retrograde approach, contralateral injection, and IVUS guidance.[7] Han successfully assessed calcified CTO lesions concluding that IVUS guidance improves the success rate in those patients.[8]

Tsujita in 2009 compared anterograde and retrograde CTO recanalization IVUS characteristics on subintimal wire tracking.[9] At that time, anterograde CTO recanalization was focused on stiffer and high torque guidewires and increased support with low profile microcatheters. Park recommended IVUS guidance for confident stiff guidewires progression.[10]

Remarkably, Rathore in 2010 proposed the intravascular ultrasound-guided reverse controlled antegrade and retrograde tracking (CART).[11] This was confirmed by Muramatsu who evaluated IVUS guiding the retrograde guidewire to its reentry site.[12]

Muhammad reviewed the incidence of subintimal guidewire tracking during successful CTO percutaneous therapy, and found that it occurs in 45% of cases[13]; this was recently confirmed by Song.[14]

Kim published an exciting example of both antegrade and retrograde subintimal tracking in a reverse RCA CTO.[15] Dai evaluated the intravascular ultrasound-guided CART in patients with unfruitful antegrade and/or retrograde CTO revascularization. The technique and procedure success rates were 95.9% and 93.9%, respectively.[16] Muramatsu assessed the incidence and impact on clinical outcomes of subintimal tracking. Multivariate analysis identified the preprocedural reference diameter as a predictor of subintimal tracking. Angiographic follow-up revealed a greater late loss in the subintimal group compared with the intimal group.[17] Dash updated the IVUS-guided wiring techniques at CTO entrance and IVUS-guided navigation on subintimal space.[18] However, in 2019 Chan reported that the CART's failure to cross the retrograde channel

with guidewire is currently the predominant CTO recanalization failure mode.[19]

Kang was concerned about high in-stent restenosis (ISR) incidence in CTO recanalization. The IVUS predictors of ISR in CTO lesions were a post-PCI minimal luminal diameter (MLD) $\leq 2.4\,\mathrm{mm}$ and a stent expansion ratio (SER) $\leq 70\%$.[20] Likewise, Mohandes identified CTO ISR as a scenario in which it is highly relevant to use IVUS to detect previous underexpansion or inadequate apposition.[21]

Scheinert was the first to evaluate an IVUS-CTO-dedicated peripheral device: the cross-point transaccess, renamed as Pioneer crossing device.[22] Another dedicated device appeared 5 years later, the Navifocus WR, which was initially evaluated by Okamura. Navifocus WR probe (similar to ViewIT) can be inserted into subintimal space (2.5 F), guiding the second guidewire to the true lumen.[23] Likewise, Kinoshita introduced the "Slipstream technique," a new concept of IVUS-guided wiring technique in the treatment of coronary total occlusions where a double-lumen IVUS catheter conducts a second guidewire.[24] More recently, Lee evaluated forward-looking (FL) IVUS.[25] Recently, Okamura evaluated the "tip detection method" using a new Navi-IVUS-based three-dimensional wiring technique for CTO intervention.[26]

Back in 2006, Lotan published the first drug eluted stent OCT study; the SICTO (sirolimus-eluting stent in CTO) study included 25 patients. IVUS was not used to guide the procedure but to evaluate intrastent proliferation at 6 months (only in 5% of cases in-stent restenosis was > to 50%). The demonstration that drug eluted stents (DES) provide a low restenosis rate was fundamental to support the long-term benefits of the CTO treatment until that time frequently questioned.[27] Hong was the first to evaluate IVUS guidance in CTO recanalization with second-generation drug-eluting stents. IVUS guidance appears to be associated with a reduction in stent thrombosis and myocardial infarction compared with angiography-guided CTO PCI. However, IVUS was not associated with a reduction in overall major cardiac adverse events (MACE).[28] It was confirmed by the AIR-CTO study in which the rates of adverse clinical events were comparable between the IVUS- and angiography-guided groups at a 2-year follow-up.[29] Controversially, Kim compared IVUS-guided CTO intervention in a randomized Zotarolimus-Eluting Versus Biolimus-Eluting Stent study. He found that IVUS guided results in 12-month major adverse cardiac events (MACE) were fewer compared with the conventional angiography-guided intervention.[30] Vaquerizo published CTO-ABSORB study, the first study with ABSORB bioresorbable everolimus-eluting scaffold in the treatment of chronic total occlusions. Because of the limited radial force and trackability, all cases were guided by IVUS and finally assessed by optical coherence tomography (OCT). IVUS facilitated the appropriate lesion preparation, permitting an adequate scaffold expansion.[31] Sabbah validated a subintimal stent deployment in recanalized CTO segments by using second-generation DES with IVUS guidance.[32] Recently, Kim conclusively demonstrated the benefit of IVUS guidance when achieving stent optimization on IVUS evaluation with new-generation DES implantation for CTO.[33]

Guo, in 2012, published the first virtual histology intravascular ultrasound analysis of coronary chronic total occlusions. CTO morphology can be divided into two patterns: (1) CTO with VH fibroatheroma and (2) CTO without VH fibroatheroma. This suggests two mechanisms of CTO formation (the majority evolving from acute coronary syndrome and thrombosis and the minority from atherosclerosis progression). The mean necrotic core/dense calcium ratio was higher in CTOs than in nonocclusive lesions (Guo in 2012). Controversially, 1 year later in a new analysis Guo published that the morphological characteristics of CTOs were similar as compared with non-CTO lesions.[34, 35] Park also found that plaque characteristics of CTO lesions were similar to non-CTO lesions.[36] Yamamoto analyzed both coronary computed tomographic angiography (CCTA) and IVUS to categorize CTO morphology.[37] Kimura combined IVUS and coronary angioscopy examination (CAS). Multivariate analysis identified bright-yellow plaque as an independent predictor of the occurrence of periprocedural myocardial necrosis. Furthermore, he found that the combination of IVUS and CAS analysis may help clarify the pathogenetic mechanism of CTO lesions.[38]

Gomez-Lara demonstrated the lumen enlargement in the distal segment of a recanalized CTO,[39] which was confirmed by Okuya,[40] and Hong.[41]

Nakashima introduced 6 Fr catheter CTO recanalization via bilateral transradial approach.[42] Sunagawa was the first to report an iatrogenic CTO after mitral valve repair.[43] Dai first reported the revascularization of complex occlusion of the right coronary artery via a gastroepiploic artery (GEA) by using IVUS-guided reverse CART technique. A soft guidewire was advanced through the right GEA collateral channel to the distal end of the CTO. In that case, the RCA was eventually revascularized by implantation of drug-eluting stents using intravascular ultrasound-guided reverse CART.[44] Takahashi initiated the transvenous IVUS-guided percutaneous coronary intervention for CTO. The procedure involved the passage of a guidewire through the CTO lesion under fluoroscopic guidance and insertion of an IVUS catheter into the cardiac vein parallel to the target artery. He reported two successful transvenous CTO revascularizations (IVUS)-guided, one with CTO of the left circumflex artery and the other with CTO of the left anterior descending artery.[45]

## 2. WHY AND WHEN IVUS IS USED IN CTO RECANALIZATION

The subintimal space has never been thoroughly studied.

1. The nature of the arterial wall includes three layers: intima, media, and adventitia. The intima layer includes the plaque and has a prominent role in atherosclerotic development. The media layer includes the smooth muscle cells conditioning the adaptive vessel diameter. This layer is sensible to chemical stimulation and local factors, and typically is relaxed with nitrites. The adventitia layer has the compliance to grow in diameter until approximately 50% in the area (20% in diameter) as positive remodeling in parallel to the plaque development. The angiographic stenosis starts when the remodeling capacity in the adventitia arrives at their limit. This is considered a slow phenomenon, but in case of spontaneous dissection, typically the adventitia shows immediate positive remodeling reaction to the lateral rejection of the media. Also, in the case of a hematoma, the adventitia shows an immediate positive remodeling reaction.

2. CTO treatment creates subintimal spaces through false lumens, which rejects laterally the true lumen, including the media. This space is identical to the ones in spontaneous dissection or caused by vessel wall hematoma and may represent a maximal remodeling with limited compliance for stenting.

IVUS provides precise online information regarding plaque composition and guidewire situation about the true lumen. Nonetheless, IVUS also allows stent selection and optimization, which reduces stent-related complications as subacute thrombosis.

Following are classifications considering the main stages where IVUS may provide support in CTO recanalization:

### 2.1. Anterograde Approach to CTO Recanalization

The main targets of IVUS in CTO anterograde recanalization approach are as follows:

1. Define the proximal edge and validate the true lumen access.

2. Delimit true lumen situation versus guidewire position along with the CTO stenosis until the true distal lumen.
3. Stent size selection and stent result optimization.

### 2.1.1. IVUS to define the proximal edge and validate the true lumen access

Sometimes it is possible to provide online real-time information. By using a parallel guidewire, the IVUS probe could be advanced into a side branch to visualize the occlusion point. In that bifurcations and ostium, the harder plaque is frequently identified in the opposite position to the side branch while the softer access point is closer to the side branch. IVUS may online monitor the guidewire to the CTO stump through the identified pathway.

It is not always possible to find an appropriate side branch for a parallel guidewire, but in any case, after an initial guidewire entrance into the CTO lesion, an early anterograde IVUS imaging is always recommended. The expansion of a small balloon with a maximal diameter of 1.5 mm is frequently required to facilitate the IVUS access through the CTO.

### 2.1.2. IVUS to delimit true lumen situation versus guidewire position along with the CTO stenosis until the true distal lumen

This early IVUS run should warranty that the guidewire is in the true lumen before active predilatation and stenting. In case that the guide is in a false lumen, IVUS may provide anterograde reentry guidance with a parallel guidewire. An attempt is made to orient the second guidewire through the true lumen, but sometimes it is only possible to re-enter the true lumen after a subendothelial false lumen segment. In any case, IVUS will help to identify procedural-related vessel damage and subintimal tracking. Finally, it is fundamental to confirm that the guidewire distal pathway is in the true lumen before performing predilatation and stenting.

In-stent CTO is a particular scenario where a suboptimal result of the previous PCI, such as incomplete stent apposition or unsatisfactory expansion, might make the procedure more challenging, justifying the use of IVUS to highlight the situation and to overcome some difficulties.

### 2.1.3. IVUS to size and optimize the stent

The use of IVUS in CTO recanalization guidance results in larger and shorter stents and reduction in geographical missing. IVUS post stenting is also recommended to verify the apposition and stent expansion, to identify geographical missing or distal dissections, and to document the final minimal luminal diameter.

## 2.2. IVUS in Retrograde CTO Recanalization

The main uses of IVUS in CTO anterograde recanalization approach are as follows:
1. IVUS in reverse recanalization
2. IVUS in reverse CART
3. IVUS in bailout
4. Size and optimize the stent.

### 2.2.1. IVUS in reverse recanalization

For retrograde CTO recanalization, IVUS is advanced through an anterograde guidewire. Most frequently, the IVUS placed as distal as possible over the anterograde guidewire. During the reverse recanalization, IVUS may confirm the intraluminal positioning when the retrograde guidewire has crossed the proximal stump. In case the CTO edge is close to a bifurcation or an ostium, the IVUS may be positioned for online imaging in the side branch.

### 2.2.2. IVUS in reverse CART

IVUS guidance is critical in the CTO technique called "controlled anterograde retrograde tracking" (CART), where a subintimal segment is proactively created. In these cases, the IVUS probe is placed as close as possible to the stump, if possible, in a side branch. This technique uses two guidewires in opposite directions, one in the anterograde direction, and the other in the retrograde direction. The IVUS may help to find the best point for dissecting the intima to connect the opposite guidewires in a circuit.

### 2.2.3. IVUS in bailout

Retrograde CTO recanalization is frequently complicated with anterograde and retrograde dissections. In case this circumstance is suspected, IVUS is strongly recommended to evaluate the dissection and guidewire position before proceeding with predilatation.1

### 2.2.4. Size and optimize the stent

Retrograde CTO recanalization may need a high-quantity contrast, so IVUS is recommended to size the stent and to verify stent expansion to save the contrast. The IVUS goal criterion is to obtain a minimal lumen area superior to 80% of the reference area covering all the stenotic plaque without significant dissections (more than type B) in the proximal edge of the stent.

# 3. CASE EXAMPLES
## 3.1. Case 1 Example
### *Anterograde RCA CTO*
#### Introduction
- Male, 60years/old, with a history of dyslipidemia and hypertensive blood pressure. LIMA to LAD 6 years ago.

- Symptoms and ECG: Angina with inferior region ischemia signs.
- Diagnostic 6 months before the intervention demonstrated a total occlusion in RCA. The patient underwent CTO recanalization program.

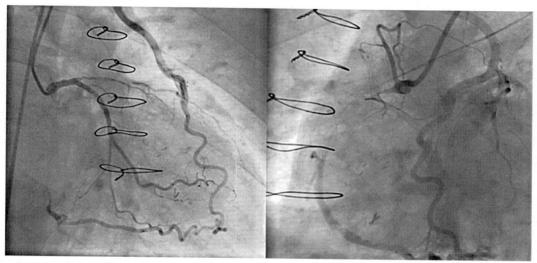

CASE 1, FIG. 1 Diagnostic angiogram confirms the presence of a CTO in right coronary artery (RCA). The distal bed of RCA receives retrograde circulation from LAD.

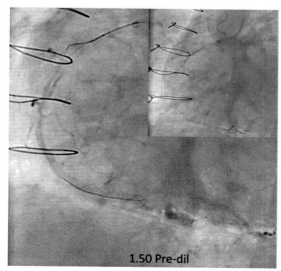

1.50 Pre-dil

CASE 1, FIG. 2 After anterograde RCA CTO guidewire passage, IVUS is used to verify the guidewire path. Sometimes, a predilatation with a small balloon is required to grant the IVUS probe access.

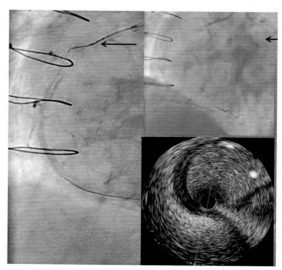

CASE 1, FIG. 3 IVUS reveals a guidewire below the media in subendothelial space.

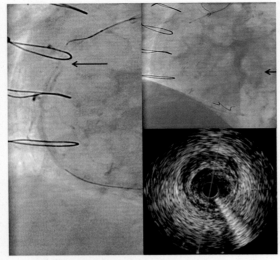

CASE 1, FIG. 4 IVUS reveals a hematoma product of subendothelial injury.

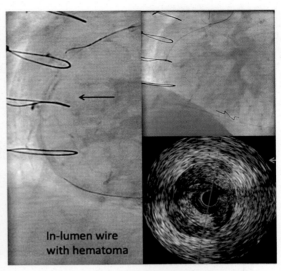

CASE 1, FIG. 6 Distally, the IVUS probe is in the true lumen, into a concentric hematoma compressing the true lumen.

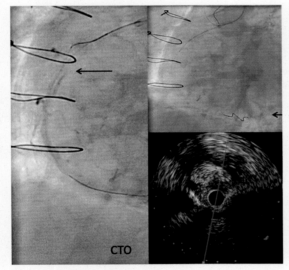

CASE 1, FIG. 5 More distally, a dissected fibro-calcific plaque.

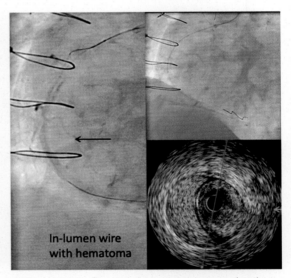

CASE 1, FIG. 7 Eccentric hematoma compressing the true lumen. Guidewires (and IVUS probe) in the true lumen.

## 3.2. Case 2 Example
*Anterograde LAD CTO*
**Introduction**

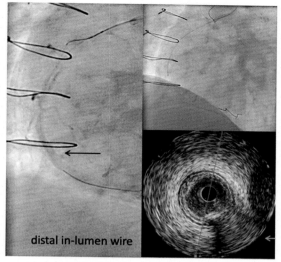

CASE 1, FIG. 8 End of the hematoma. Guidewire (and IVUS probe) in the true lumen.

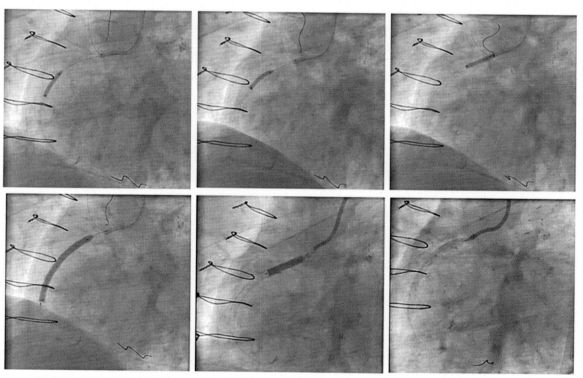

CASE 1, FIG. 9 Despite the complex case, the vessel can be safely stented.

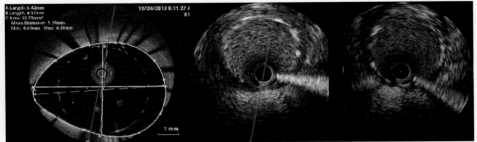

**Proximal incomplete apposition 1mm length**

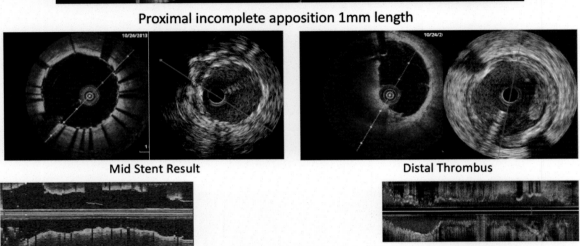

**Mid Stent Result**     **Distal Thrombus**

CASE 1, FIG. 10 Poststent IVUS and OCT assessment demonstrate a short segment of incomplete apposition and a small thrombus at the distal stent edge.

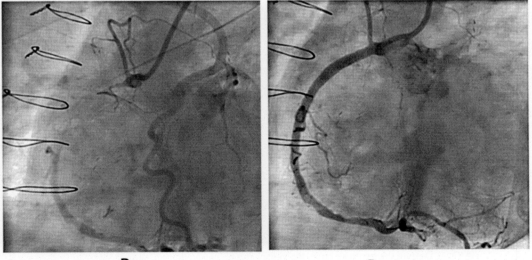

**Pre**     **Post**

CASE 1, FIG. 11 Basal and final angiograms.

- Male, 51 years/old, with a history of hypertensive blood pressure.
- Antecedents of a stent in left main 1 year ago.
- Symptoms and ECG: Anterior-lateral heart region ischemia with angina.

- Diagnostic 6 months before the intervention demonstrated a total occlusion in the left anterior descendent (LAD). The patient underwent CTO recanalization program.

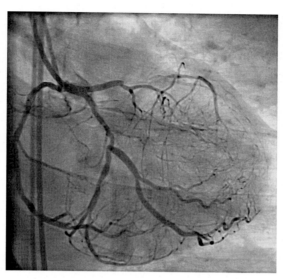

CASE 2, FIG. 1 Diagnostic angiogram with double catheterization (right coronary artery and left coronary tree) confirms the presence of a CTO in LAD after the first diagonal. The distal bed of LAD receives some collateral circulation.

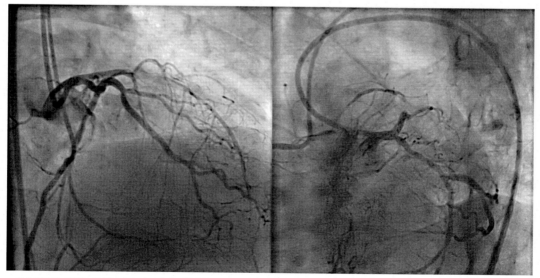

CASE 2, FIG. 2 Best projections (RAO and Spider) that were identified.

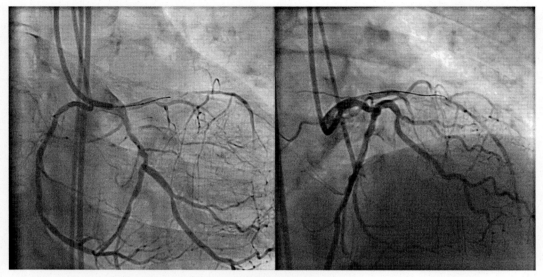

CASE 2, FIG. 3 Guidewire cross CTO by angiographic monitoring.

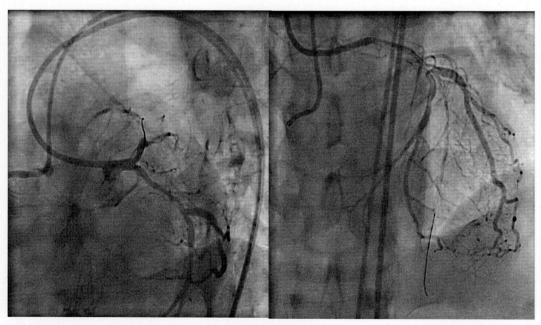

CASE 2, FIG. 4 Guidewire is introduced without difficulties.

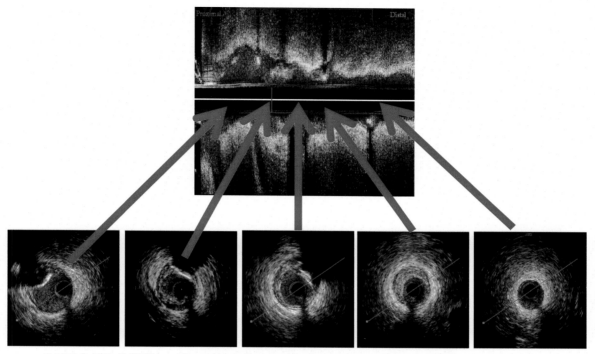

CASE 2, FIG. 5 IVUS is performed to verify the guidewire pathway, enlightening of a plaque dissection and indicating that the guidewire is into the true lumen.

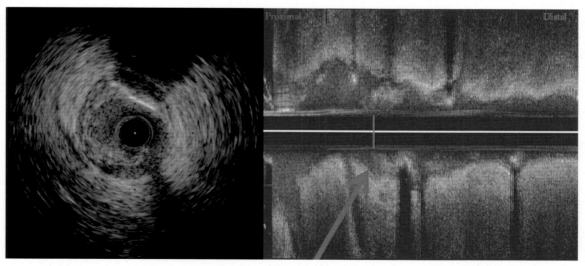

CASE 2, FIG. 6 AS IVUS confirmed the guidewire is in the true lumen, plaque dissection ought to be stented.

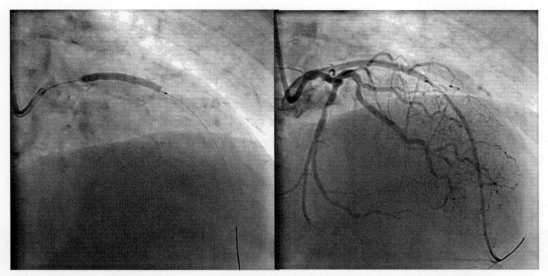

CASE 2, FIG. 7 After predilatation and stenting, satisfactory angiographic result was displayed.

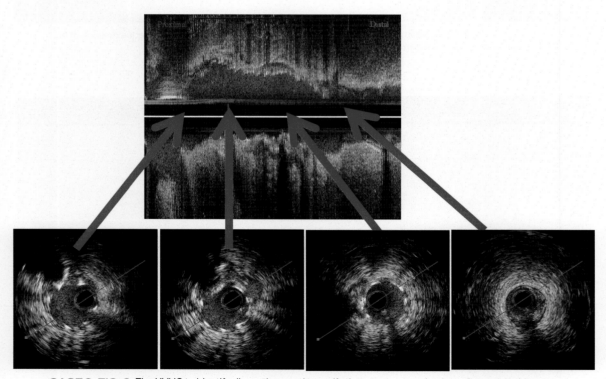

CASE 2, FIG. 8 Final IVUS to identify dissections and to verify that stent expansion is performed. In this case, IVUS confirmed excellent result.

## 3.3. Case 3 Example
*Ostial LAD CTO procedure guided
from diagonal IVUS*
### Introduction
- Male, 47 years/old, with a history of hypertensive blood pressure.
- No other antecedents.
- Symptoms and ECG: Angina with anterior ischemia.
- Diagnostic: The patient underwent CTO recanalization program.

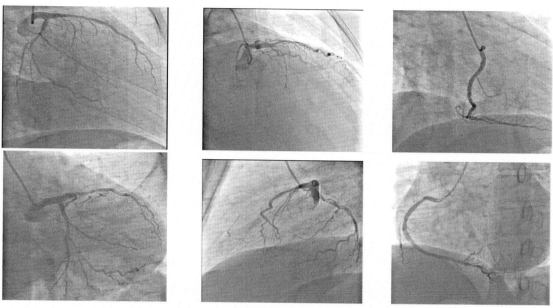

CASE 3, FIG. 1 Diagnostic angiogram confirms the presence of a CTO in LAD at the level first diagonal.

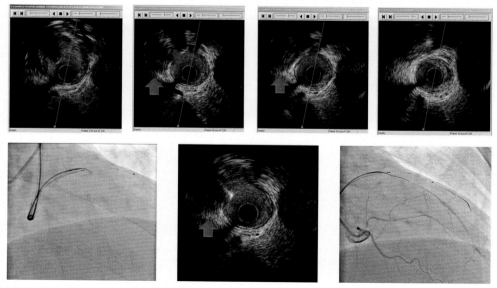

CASE 3, FIG. 2 A first guidewire was placed in the first diagonal to monitor the CTO procedure with a second guidewire in LAD. The LAD stump was clearly visible from the diagonal.

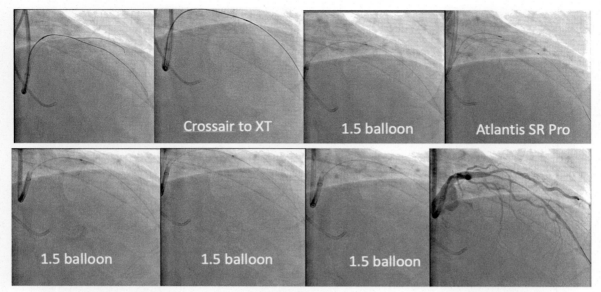

CASE 3, FIG. 3 Recanalization was possible with a Conquest Pro guidewire. Then, a 1.50-mm predilatation was performed.

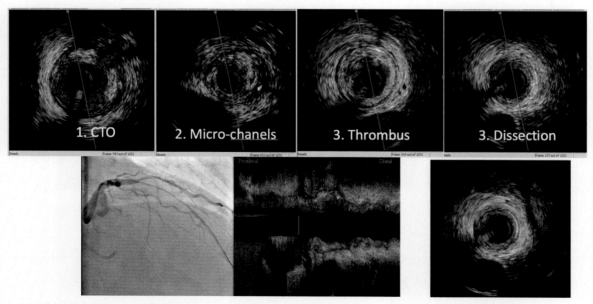

CASE 3, FIG. 4 Inside the LAD, different plaque characteristics and references were identified by IVUS, confirming the true lumen guidewire position.

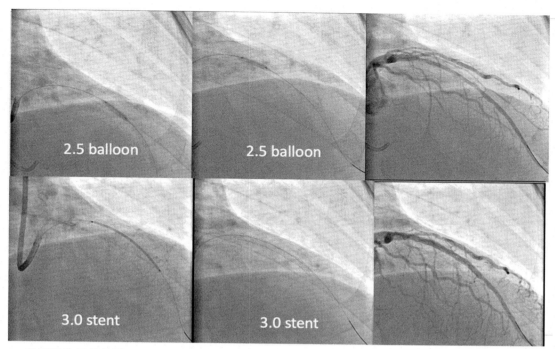

CASE 3, FIG. 5 The artery was dilated and stented.

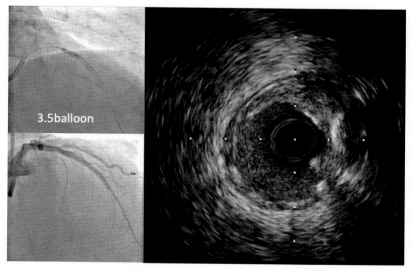

CASE 3, FIG. 6 Stent postdilatation was performed. The result was evaluated by IVUS, confirming stent expansion without any significant dissection.

## 3.4. Case 4 Example
*Calcified CTO in LAD*
### Introduction
- Male, 62 years/old.

- As antecedents, a previous stent implanted in circumflex (CX).
- Symptoms and ECG: Anterior region ischemia with angina.

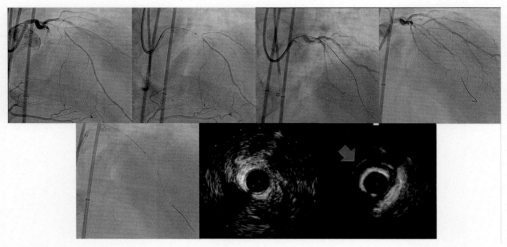

CASE 4, FIG. 1 After a double catheterization, angiograms confirm the presence of CTO in LAD, with retrograde flow from right coronary artery (RCA). After crossing CTO with the guidewire, IVUS reveals a concentric calcification located proximal to the CTO origin and rotational atherectomy was recommended.

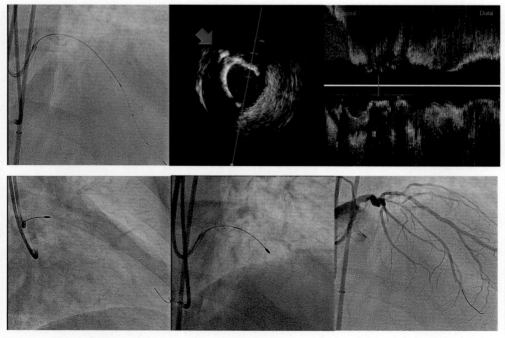

CASE 4, FIG. 2 Rotational atherectomy was performed with great improvement.

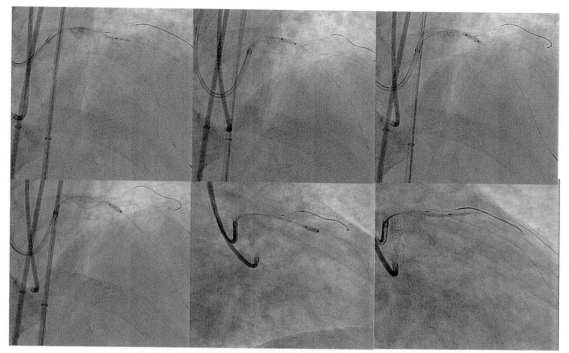

CASE 4, FIG. 3 CTO lesion was dilated with balloons.

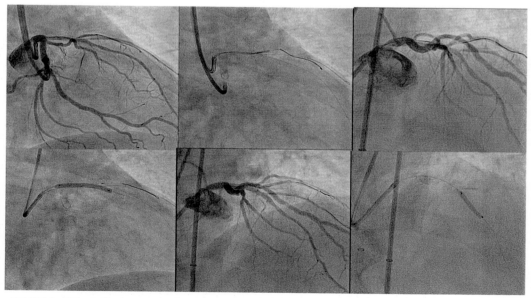

CASE 4, FIG. 4 Stents were implanted, and balloon postdilatation was performed.

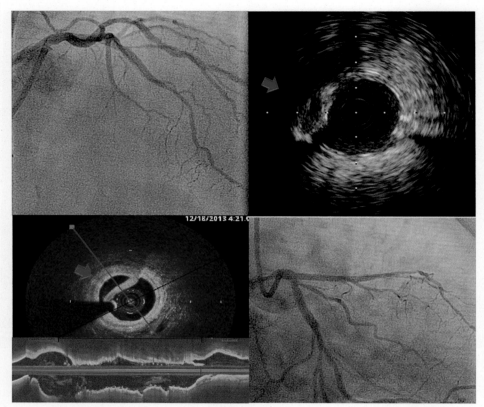

CASE 4, FIG. 5  A distal dissection was identified by IVUS and confirmed with OCT. An additional stent was implanted with excellent result.

- Diagnostic: The patient underwent CTO recanalization program because CTO in LAD artery was confirmed.

### 3.5. Case 5 Example
*LAD CTO, anterograde wired in false lumen*
**Introduction**

- Male, 68 years/old.
- Symptoms and ECG: Angina with antero-lateral region ischemia.
- Diagnostic: The patient underwent CTO recanalization program because CTO in LAD artery.

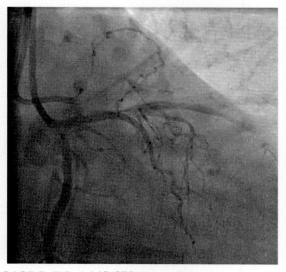

CASE 5, FIG. 1  LAD CTO was confirmed.

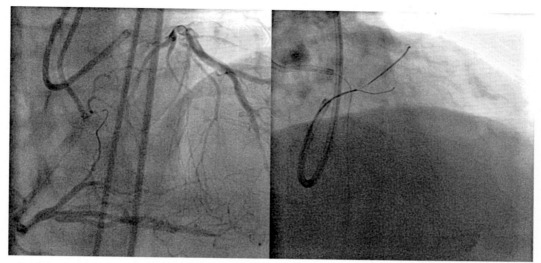

CASE 5, FIG. 2 LAD recanalization was attempted with difficulties to advance the guidewire. Simultaneous RCA catheterization gives angiographic information on the distal vessel. However, anterograde catheterization was difficult and needed different guidewires attempts.

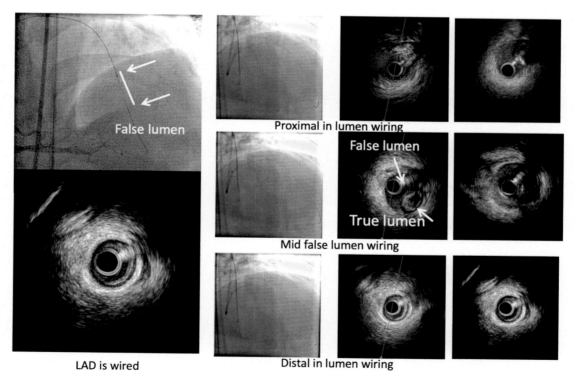

CASE 5, FIG. 3 When the guidewire finally was placed, IVUS demonstrated a subintimal guidewire position with a long segment in false lumen.

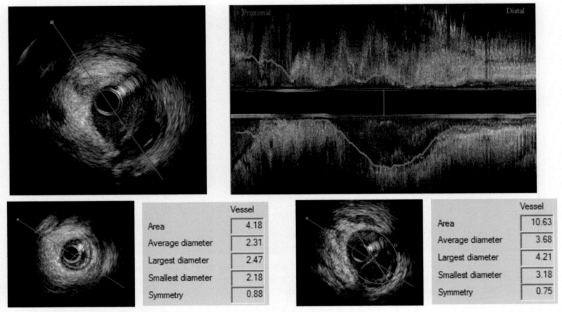

CASE 5, FIG. 4 IVUS longitudinal view displays the subintimal segment.

| | Vessel |
|---|---|
| Area | 4.18 |
| Average diameter | 2.31 |
| Largest diameter | 2.47 |
| Smallest diameter | 2.18 |
| Symmetry | 0.88 |

| | Vessel |
|---|---|
| Area | 10.63 |
| Average diameter | 3.68 |
| Largest diameter | 4.21 |
| Smallest diameter | 3.18 |
| Symmetry | 0.75 |

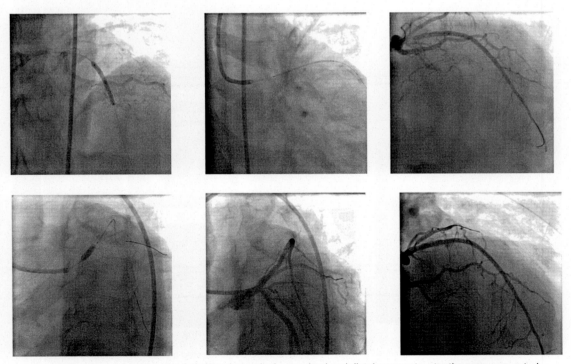

CASE 5, FIG. 5 As guidewire is the true lumen in the proximal and distal segments, stenting was suggested.

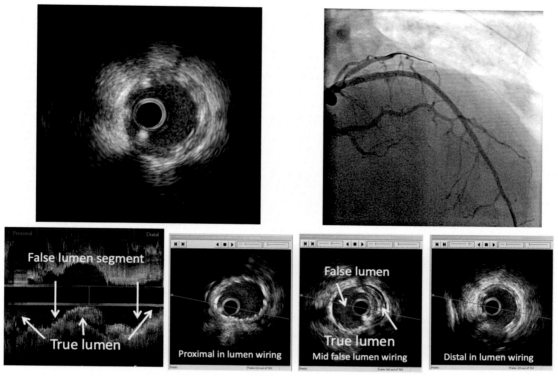

CASE 5, FIG. 6 IVUS confirms the good angiographic results.

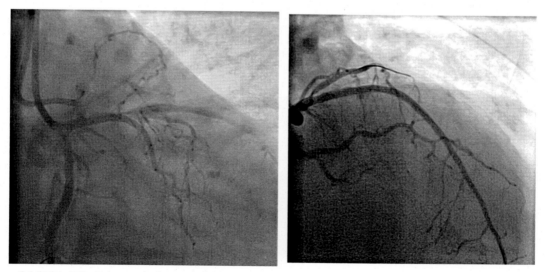

CASE 5, FIG. 7 Images before and after angiograms.

### 3.6. Case 6 Example
*LM severe stenosis and LAD CTO*
#### Introduction
- Male, 63 years/old.
- Symptoms and ECG: Severe angina with chest pain and syncope. Antero-lateral region ischemia was documented.

- Diagnostic: The patient underwent urgent recanalization because syncope with severe antero-lateral ischemia.

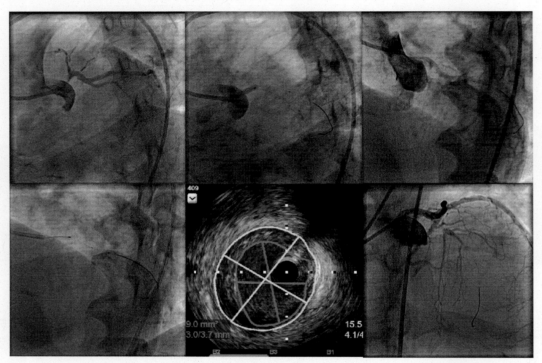

CASE 6, FIG. 1 Patient presented chest pain and hemodynamic instability after guiding catheter was positioned. Diagnostic angiogram showed severe LM stenosis and CTO in proximal LAD, just after a first diagonal branch. A balloon inflation was performed in LM to improve patient perfusion and an IVUS was performed in LM showing a normalized lumen diameter.

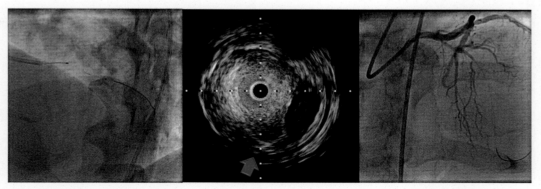

CASE 6, FIG. 2 With the guidewire placed in the remaining diagonal, IVUS identified the LAD CTO stump, guiding the LAD wire pathway.

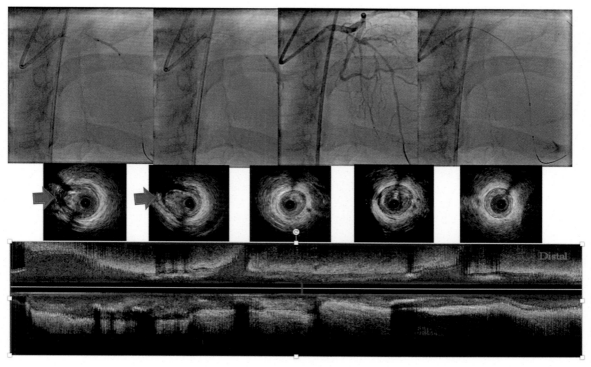

CASE 6, FIG. 3 With the guidewire placed in the open diagonal, we identified the LAD CTO stump, guiding the LAD wire pathway.

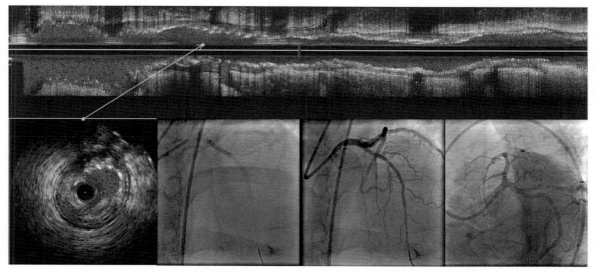

CASE 6, FIG. 4 Stent was implanted from LAD to LM with excellent result verified with IVUS.

### 3.7. Case 7 Example
*LAD CTO, bailout retrograde from RCA*
**Introduction**
- Male, 65 years/old.
- Symptoms and ECG: Severe angina with an antero-lateral region ischemia.

- Diagnostic: The patient underwent CTO recanalization program because of the demonstration of CTO in LAD artery.

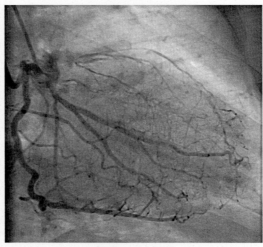

CASE 7, FIG. 1 With double catheterization, some retrograde flow from RCA to LAD was demonstrated.

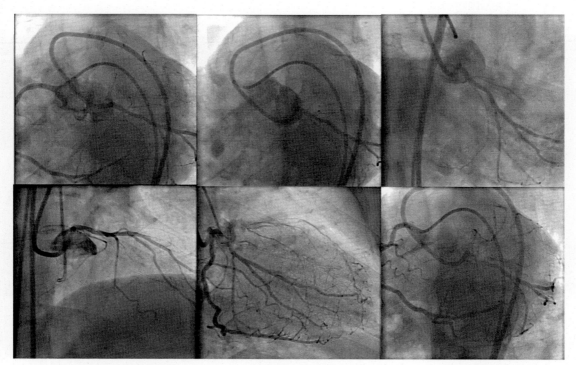

CASE 7, FIG. 2 Below the basal angiograms. Stumpless CTO lesion became very challenging in anterograde approach.

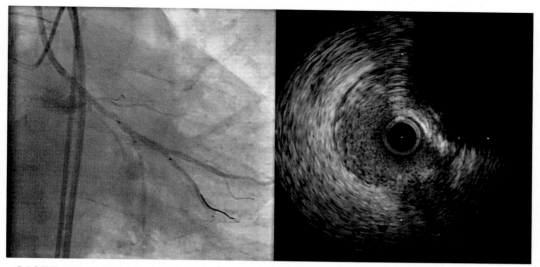

CASE 7, FIG. 3 Anterograde approach fail after multiple attempts. However, a long LAD segment is wired in the true lumen as confirmed by IVUS.

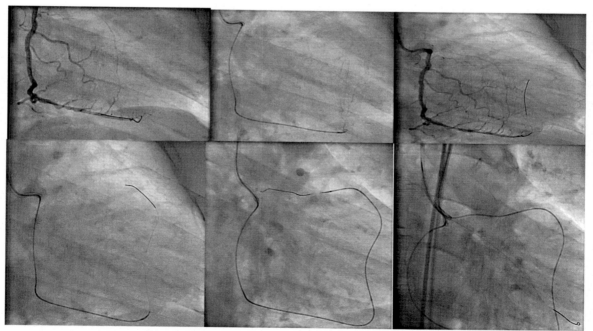

CASE 7, FIG. 4 RCA is wired to find a good septal branch for retrograde LAD access. Retrograde recanalization was successful with full RCA-septal-LAD loop guidewire externalization through the left guiding catheter.

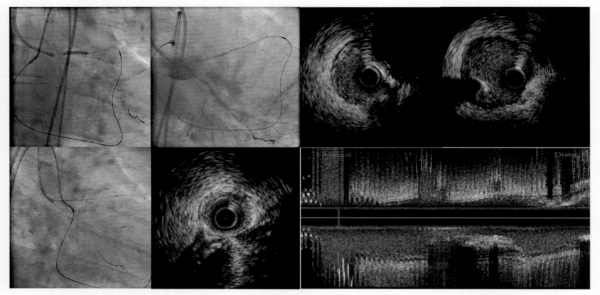

CASE 7, FIG. 5 Anterograde IVUS demonstrated the LAD true lumen retrograde guidewire position.

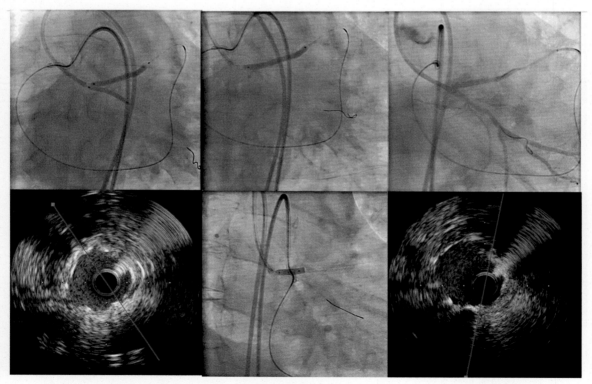

CASE 7, FIG. 6 Predilatation, stent implantation, and IVUS control. Stent infra-expansion was identified, and postdilatation was recommended. Adequate final IVUS result.

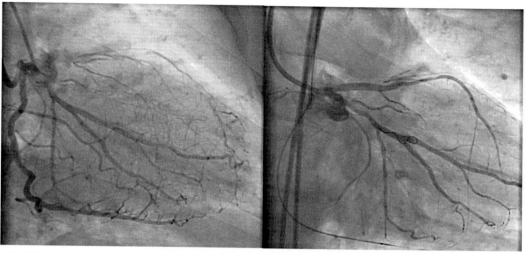

CASE 7, FIG. 7 Basal preprocedural and final postprocedural images comparison.

# REFERENCES

1. Holmes DR, Vlietstra RE, Reeder GS, et al. Angioplasty in total coronary artery occlusion. *J Am Coll Cardiol.* 1984; 3:845–849. https://doi.org/10.1016/s0735-1097(84)80263-6.
2. Serruys PW, Umans V, Heyndrickx GR, et al. Elective PTCA of totally occluded coronary arteries not associated with acute myocardial infarction; short-term and long-term results. *Eur Heart J.* 1985; 6:2–12. https://doi.org/10.1093/oxfordjournals.eurheartj.a061750.
3. Kimura BJ, Tsimikas S, Bhargava V, DeMaria AN, Penny WF. Subintimal wire position during angioplasty of a chronic total coronary occlusion: detection and subsequent procedural guidance by intravascular ultrasound. *Catheter Cardiovasc Diagn.* 1995; 35(3):262–265.
4. Matsubara T, Murata A, Kanyama H, Ogino A. IVUS-guided wiring technique: promising approach for the chronic total occlusion. *Catheter Cardiovasc Interv.* 2004; 61(3):381–386.
5. Ozawa N. A new understanding of chronic total occlusion from a novel PCI technique that involves a retrograde approach to the right coronary artery via a septal branch and passing of the guidewire to a guiding catheter on the other side of the lesion. *Catheter Cardiovasc Interv.* 2006; 68(6):907–913.
6. García-García HM, Kukreja N, Daemen J, et al. Contemporary treatment of patients with chronic total occlusion: critical appraisal of different state-of-the-art techniques and devices. *EuroIntervention.* 2007; 3(2):188–196.
7. Cuneo A, Tebbe U. The management of chronic total coronary occlusions. *Minerva Cardioangiol.* 2008; 56(5): 527–541.
8. Han YL, Zhao HQ, Wang SL, et al. Acute angiographic and clinical outcomes of patients with calcified chronic total occlusion underwent percutaneous coronary intervention. *Zhonghua Xin Xue Guan Bing Za Zhi.* 2008; 36(8):706–709.
9. Tsujita K, Maehara A, Mintz GS, et al. Intravascular ultrasound comparison of the retrograde versus antegrade approach to percutaneous intervention for chronic total coronary occlusions. *JACC Cardiovasc Interv.* 2009; 2(9):846–854.
10. Park Y, Park HS, Jang GL, et al. Intravascular ultrasound guided recanalization of stumpless chronic total occlusion. *Int J Cardiol.* 2011; 148(2):174–178.
11. Rathore S, Katoh O, Tuschikane E, Oida A, Suzuki T, Takase S. A novel modification of the retrograde approach for the recanalization of chronic total occlusion of the coronary arteries intravascular ultrasound-guided reverse controlled antegrade and retrograde tracking. *JACC Cardiovasc Interv.* 2010; 3(2):155–164.
12. Muramatsu T, Tsukahara R, Ito Y. A novel intravascular ultrasound-guided percutaneous coronary angioplasty technique via the retrograde approach for chronic total occlusion. *Cardiovasc Interv Ther.* 2011; 26(1):45–51.
13. Muhammad KI, Lombardi WL, Christofferson R, Whitlow PL. Subintimal guidewire tracking during successful percutaneous therapy for chronic coronary total occlusions: insights from an intravascular ultrasound analysis. *Catheter Cardiovasc Interv.* 2012; 79(1):43–48.
14. Song L, Maehara A, Finn MT, et al. Intravascular ultrasound analysis of intraplaque versus subintimal tracking in percutaneous intervention for coronary chronic total occlusions and association with procedural outcomes. *JACC Cardiovasc Interv.* 2017; 10:1011–1021.

15. Kim YH, Hwang SH, Lim CH, et al. Reverse controlled antegrade and retrograde subintimal tracking in chronic total occlusion of right coronary artery. *Korean Circ J.* 2012; 42(9):625–628.

16. Dai J, Katoh O, Kyo E, Tsuji T, Watanabe S, Ohya H. Approach for chronic total occlusion with intravascular ultrasound-guided reverse controlled antegrade and retrograde tracking technique: single center experience. *J Interv Cardiol.* 2013; 26(5):434–443.

17. Muramatsu T, Tsuchikane E, Oikawa Y, et al. Incidence and impact on midterm outcome of controlled subintimal tracking in patients with successful recanalisation of chronic total occlusions: J-PROCTOR registry. *Euro Intervention.* 2014; 10(6):681–688.

18. Dash D, Li L. Intravascular ultrasound guided percutaneous coronary intervention for chronic total occlusion. *Curr Cardiol Rev.* 2015; 11(4):323–327.

19. Chan CY, Wu EB, Yan BP, Tsuchikane E. Procedure failure of chronic total occlusion percutaneous coronary intervention in an algorithm driven contemporary Asia-Pacific Chronic Total Occlusion Club (APCTO Club) multicenter registry. *Catheter Cardiovasc Interv.* 2019; 93(6): 1033–1038.

20. Kang J, Cho YS, Kim SW, et al. Intravascular ultrasound and angiographic predictors of in-stent restenosis of chronic total occlusion lesions. *PLoS ONE.* 2015; 10. e0140421.

21. Mohandes M, Vinhas H, Fernández F, Moreno C, Torres M, Guarinos J. When intravascular ultrasound becomes indispensable in percutaneous coronary intervention of a chronic total occlusion. *Cardiovasc Revasc Med.* 2018; 19(3 Pt A):292–297.

22. Scheinert D, Bräunlich S, Scheinert S, Ulrich M, Biamino G, Schmidt A. Initial clinical experience with an IVUS-guided transmembrane puncture device to facilitate recanalization of total femoral artery occlusions. *EuroIntervention.* 2005; 1:115–119.

23. Okamura A, Iwakura K, Fujii K. ViewIT improves intravascular ultrasound-guided wiring in coronary intervention of chronic total occlusion. *Catheter Cardiovasc Interv.* 2010; 75(7):1062–1066.

24. Kinoshita Y, Fujiwara H, Suzuki T. "Slipstream technique"—new concept of intravascular ultrasound guided wiring technique with double lumen catheter in the treatment of coronary total occlusions. *J Cardiol Cases.* 2017; 16(2):52–55.

25. Lee J, Chang JH. A 40-MHz ultrasound transducer with an angled aperture for guiding percutaneous revascularization of chronic total occlusion: a feasibility study. *Sensors (Basel).* 2018; 18(11):4079.

26. Okamura A, Iwakura K, Iwamoto M, et al. Tip detection method using the new IVUS facilitates the 3-dimensional wiring technique for CTO intervention. *JACC Cardiovasc Interv.* 2019; https://doi.org/10.1016/j.jcin.2019.07.041.

27. Lotan C, Almagor Y, Kuiper K, Suttorp MJ, Wijns W. Sirolimus-eluting stent in chronic total occlusion: the SICTO study. *J Interv Cardiol.* 2006; 19(4):307–312.

28. Hong SJ, Kim BK, Shin DH, et al. Usefulness of intravascular ultrasound guidance in percutaneous coronary intervention with second-generation drug-eluting stents for chronic total occlusions (from the Multicenter Korean-Chronic Total Occlusion Registry). *Am J Cardiol.* 2014; 114(4):534–540.

29. Tian NL, Gami SK, Ye F, et al. Angiographic and clinical comparisons of intravascular ultrasound- versus angiography-guided drug-eluting stent implantation for patients with chronic total occlusion lesions: two-year results from a randomised AIR-CTO study. *EuroIntervention.* 2015; 10(12):1409–1417.

30. Kim BK, Shin DH, Hong MK, et al. Clinical impact of intravascular ultrasound-guided chronic total occlusion intervention with zotarolimus-eluting versus biolimus-eluting stent implantation: randomized study. *Circ Cardiovasc Interv.* 2015; 8(7). e002592.

31. Vaquerizo B, Barros A, Pujadas S, et al. Bioresorbable everolimus-eluting vascular scaffold for the treatment of chronic total occlusions: CTO-ABSORB pilot study. *Euro Intervention.* 2015; 11(5):555–563.

32. Sabbah M, Tada T, Kadota K, et al. Clinical and angiographic outcomes of true vs. false lumen stenting of coronary chronic total occlusions: insights from intravascular ultrasound. *Catheter Cardiovasc Interv.* 2019; 93(3):E120–E129.

33. Kim D, Hong SJ, Kim BK, et al. Outcomes of stent optimisation in intravascular ultrasound-guided intervention for long or chronic totally occluded coronary lesions. *EuroIntervention.* 2019; https://doi.org/10.4244/EIJ-D-19-00762.

34. Guo J, Maehara A, Mintz GS, et al. A virtual histology intravascular ultrasound analysis of coronary chronic total occlusions. *Catheter Cardiovasc Interv.* 2013; 81(3):464–470.

35. Guo J, Maehara A, Guo N, et al. Virtual histology intravascular ultrasound comparison of coronary chronic total occlusions versus non-occlusive lesions. *Int J Card Imaging.* 2013; 29(6):1249–1254.

36. Park YH, Kim YK, Seo DJ, et al. Analysis of plaque composition in coronary chronic total occlusion lesion using virtual histology-intravascular ultrasound. *Korean Circ J.* 2016; 46(1):33–40.

37. Yamamoto MH, Maehara A, Poon M, et al. Morphological assessment of chronic total occlusions by combined coronary computed tomographic angiography and intravascular ultrasound imaging. *Eur Heart J Cardiovasc Imaging.* 2017; 18(3):315–322.

38. Kimura S, Sugiyama T, Hishikari K, et al. Intravascular ultrasound and angioscopy assessment of coronary plaque components in chronic totally occluded lesions. *Circ J.* 2018; 82(8):2032–2040.

39. Gomez-Lara J, Teruel L, Homs S, et al. Lumen enlargement of the coronary segments located distal to chronic total occlusions successfully treated with drug-eluting stents at follow-up. *EuroIntervention.* 2014; 9(10): 1181–1188.

40. Okuya Y, Saito Y, Takahashi T, Kishi K, Hiasa Y. Novel predictors of late lumen enlargement in distal reference segments after successful recanalization of coronary chronic total occlusion. *Catheter Cardiovasc Interv.* 2019; 94:546–552. https://doi.org/10.1002/ccd.28143.

41. Hong SJ, Kim BK, Kim YJ, et al. Incidence, predictors, and outcomes of distal vessel expansion on follow-up intravascular ultrasound after recanalization of chronic total occlusions using new-generation drug-eluting stents: data from the CTO-IVUS randomized trial. *Catheter Cardiovasc Interv.* 2019; https://doi.org/10.1002/ccd.28461.

42. Nakashima M, Ikari Y, Aoki J, Tanabe K, Tanimoto S, Hara K. Intravascular ultrasound-guided chronic total occlusion wiring technique using 6 Fr catheters via bilateral transradial approach. *Cardiovasc Interv Ther.* 2015; 30(1):68–71.

43. Sunagawa O, Nakamura M, Hokama R, Miyara T, Taba Y, Touma T. A case of percutaneous coronary intervention for treatment of iatrogenic chronic total occlusion of the left circumflex artery after mitral valve repair. *Cardiovasc Interv Ther.* 2017; 32(2):146–150.

44. Dai J, Katoh O, Zhou H, Kyo E. First reported revascularization of complex occlusion of the right coronary artery using the IVUS-guided reverse CART technique via a gastroepiploic artery graft. *Heart Vessel.* 2016; 31(2):251–255.

45. Takahashi Y, Okazaki H, Mizuno K. Transvenous IVUS-guided percutaneous coronary intervention for chronic total occlusion: a novel strategy. *J Invasive Cardiol.* 2013; 25(7):E143–E146.

# CHAPTER 4

# Intracardiac Ultrasound

JURGEN M.R. LIGTHART • KAREN TH. WITBERG
Erasmus MC, Rotterdam, The Netherlands

## 1. INTRODUCTION

Intracardiac ultrasound (ICE) was introduced as a substitute for transoesophageal echocardiography (TEE)[1, 2] to guide percutaneous interventions like transseptal punctures, patent foramen ovale (PFO) and atrial-septum defects (ASD) closures, left atrial appendage closures, and alcohol ablations for hypertrophic cardiomyopathy. Several manufacturers have introduced systems with catheters either with a rotating probe or with phased array probes. The catheters were designed to be introduced in the femoral vein to reach the right atrium or vena cava superior, from where the heart structures were scanned.

ICE has certain similarities with TEE, as well as advances and limitations.

Similarities:
- higher ultrasound frequencies and enhanced special resolution (6–12 MHz)
- anatomic assessment
- hemodynamic assessment (only phased array systems)
- guidance and monitoring of percutaneous procedures
- improved imaging in obese patients with emphysema compared to transthoracic echo.

Advances of ICE compared to TEE:
- no additional discomfort for the patient
- no general anesthesia and intubation are necessary in supine position
- not contraindicated for patients with a history of dysphagia or esophageal disease
- no artifacts due to air within the esophagus, the stomach, or the trachea

Limitations of ICE compared to TEE:
- require an 8–10F sheath venous access;
- only to be used in a catheterization laboratory by an interventional cardiologist, not in outpatient settings
- limited sector imaging (however, work is in progress with 3D imaging in phased array systems)

- no harmonics
- only single use
- especially for phased array systems, there is a learning curve for interventional cardiologists for positioning (and for orientation)

Intracardiac ultrasound is indicated to guide percutaneous procedures[3]:
- transseptal punctures
- percutaneous closure of PFO or ASD
- PV ablation (electrophysiology)
- TAVI
- left atrial appendage closure
- septal ablation for hypertrophic obstructive cardiomyopathy (HOCM)

## 2. ICE DEVICES AND CASE STUDIES

The picture gallery and the case examples in the following paragraphs illustrate the use of intracardiac ultrasound in procedure guidance. Techniques are developing fast nowadays and it is well possible that while this chapter is in press new features in the field of intracardiac ultrasound were announced or available. Harmonic echo may be implemented in intracardiac ultrasound to improve the image quality in general. Developments in real-time 3D/4D imaging like a wider sector angle (>22 degrees) may improve the possibilities to guide percutaneous valve implantations with intracardiac ultrasound. Minimizing the size of the echo consoles while maintaining all features or even full integration in intervention labs may improve the ergonomics of ICE-guided procedures in the Cath lab.

## 3. ROTATIONAL SYSTEM

The Boston Scientific Ultra ICE Plus, shown in this paragraph, has a single 9 MHz transducer inside the catheter, rotated with 1800 rpm by a rotating drive shaft, producing a 360 degrees "cross-sectional" image of the heart structure (Figs. 1–6).

**Intravascular Ultrasound.** https://doi.org/10.1016/B978-0-12-818833-0.00004-7

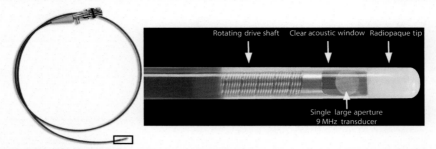

FIG. 1 Ultra ICE Plus Imaging Catheter (Boston Scientific), 9F rotational system. This catheter has no guidewire port or other build in steering possibilities and has to be introduced in the femoral vein with a long 9F steerable introducer sheath. By changing the direction of the tip of the introducer sheath, the probe can be directed to the required position. ((Right image courtesy of Boston Scientific. ©2020 Boston Scientific Corporation or its affiliates. All rights reserved.))

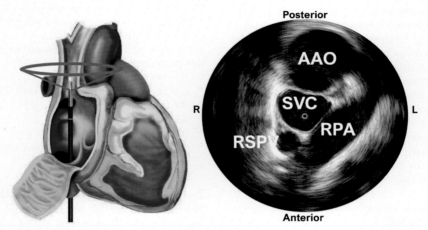

FIG. 2 Rotational catheter positioned in superior vena cava (SVC). The cross-sectional view shows ascending aorta (AAO), right pulmonary artery (RPA), and right superior pulmonary vein (RSPV). The catheter tip is visible in the SVC. ((Image courtesy of Boston Scientific. ©2020 Boston Scientific Corporation or its affiliates. All rights reserved.))

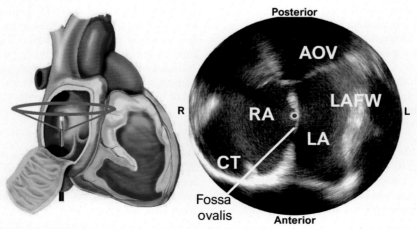

FIG. 3 Rotational catheter positioned in right atrium (RA).[4] The cross-sectional view shows crista terminalis (CT), left atrium (LA) with its free wall (LAFW), and aortic valve (AOV). The *line* indicates fossa ovalis. ((Image courtesy of Boston Scientific. ©2020 Boston Scientific Corporation or its affiliates. All rights reserved.))

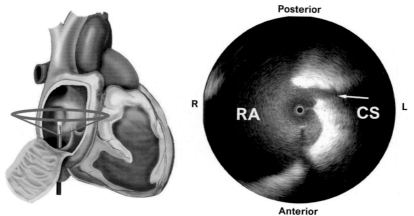

FIG. 4 Rotational catheter positioned at the floor of the right atrium (RA). The cross-sectional view shows the coronary sinus (CS) comes out into the right atrium. ((Image courtesy of Boston Scientific. ©2020 Boston Scientific Corporation or its affiliates. All rights reserved.))

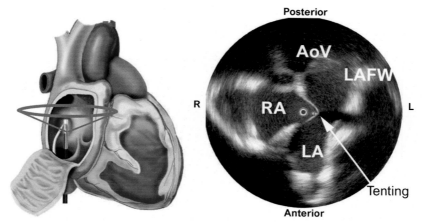

FIG. 5 Intracardiac guidance of an atrial septum puncture.[5] Rotational catheter positioned in right atrium (RA) visualizing the left atrium (LA) with its free wall (LAFW) and aortic valve (AOV). The *line* indicates the tenting by needle puncturing the septum at the fossa ovalis. ((Image courtesy of Boston Scientific. ©2020 Boston Scientific Corporation or its affiliates. All rights reserved.))

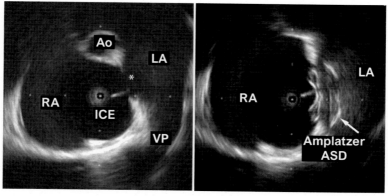

FIG. 6 Case example of a rotational ICE-guided closure of an atrial septum defect.[6] The *left image* shows the intra-atrial wall defect (*). The *right image* shows the Amplatzer ASD device implanted in the septum closing the defect. ((From Hernandez F, et al. [Intracardiac echocardiography and percutaneous closure of atrial septal defects in adults] Ecocardiografia intracardiaca en el cierre percutaneo de defectos del septo interauricular en adultos. *Rev Esp Cardiol*, 2008;61(5):465-70.))

## 4. PHASED ARRAY SYSTEM

A phased array system like Siemens Acuson Acunav (Fig. 7) provides 2D long-axis images in a 90 degrees sector with working frequencies of 5.5, 7.5, 8.5 and 10 MHz; frequencies depend on the console used. Besides, the phased array system provides color Doppler flow images and hemodynamic measurement with pulse Doppler. For operators, who routinely use TEE systems, there may be a learning curve to become familiar with interpretation of the images. Although the images have a strong resemblance with TEE, the different positions of the Intracardiac probe (mostly from inside the right atrium) compared to the position of the TEE probe (the esophagus outside the heart) may cause confusion about the image layout.

Another learning curve for the interventional cardiologist may be the positioning of the intracardiac ultrasound catheter itself. Apart from rotation, the phased array catheter can be controlled by two rotating knobs on the shaft, moving the probe on the tip to anterior and posterior and to left and right. This flexibility is a strong feature of a phased array system; however, small movements of the catheter cause immediate changes in the image. It is advisable to manipulate these catheters with "slow" movements to get the best results.

## 5. POSITION AND IMAGE EXAMPLES OF PHASED ARRAY SYSTEM[3]

The following examples show four basic positions of the intracardiac probe and the resulting images (Figs. 8–11).

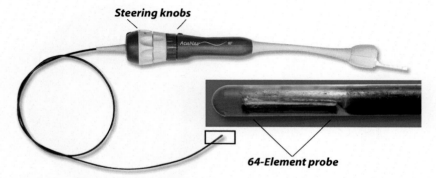

FIG. 7 Siemens Acuson Acunav phased array intracardiac ultrasound catheter. On *top*: tip of the catheter with the 64-element probe. *Bottom*: overview of the 90-cm long catheter with rotating knobs for posterior-anterior and left-right movements. ((NB: photograph taken from a used catheter.))

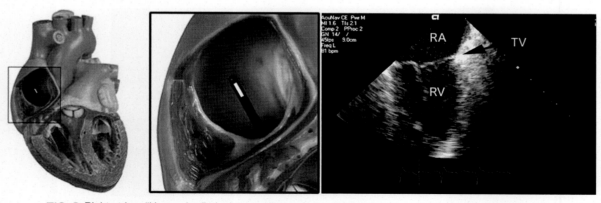

FIG. 8 Right atrium "Home view": the intracardiac probe is positioned in the mid-right atrium (RA) and faces toward the tricuspid valve (TV) and the right ventricle (RV).

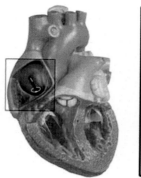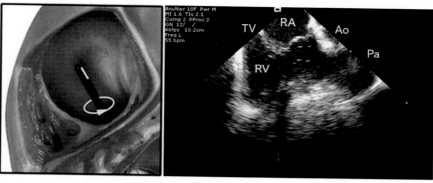

FIG. 9 Right atrium "Home view" and rotating the catheter clockwise: probe is positioned in the mid-right atrium (RA) and faces toward the tricuspid valve (TV) and the right ventricle (RV). At the right side, the aorta (Ao) and the pulmonary artery (Pa) are visible.

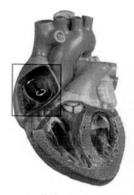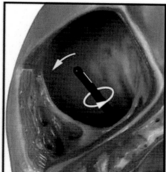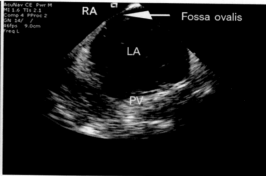

FIG. 10 Right atrium "Home view," rotating the catheter clockwise and tilting the catheter to posterior: probe is positioned in the mid-right atrium (RA) and faces toward the atrial septum with fossa ovalis and left atrium (LA). One of the pulmonary veins (PV) is visible as a "cutout."

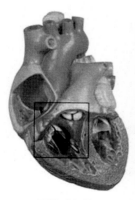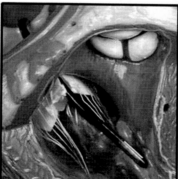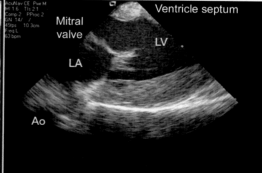

FIG. 11 From right atrium "Home view" the probe can be positioned in the right ventricle by tilting the catheter anterior toward the tricuspid valve and forwarding it into the right ventricle. The probe is facing the ventricle septum, left ventricle (LV), mitral valve, and left atrium (LA) outflow track. The descending aorta (Ao) is just visible.

## 6. CASE EXAMPLES

Intracardiac ultrasound is mainly used to guide percutaneous procedures. The following paragraphs show examples of various procedures guided with phased array intracardiac ultrasound. Justification of indications of the percutaneous procedures is beyond the scope of this chapter although the authors realize that there are contradictions in literature about, for example, closure of a PFO.

## 7. PATENT FORAMEN OVALE[7]

To guide a percutaneous PFO closure, the intracardiac ultrasound catheter is positioned high in the right atrium with the tip tilted toward posterio (Fig. 12).

## 8. ATRIAL SEPTUM DEFECT

The guidance of a percutaneous closure of an ASD can be for a 2D intracardiac ultrasound system more challenging. Different from PFO the defect of the atrial septum can be located anywhere in the septum and it may take some more time to find the correct position for the ultrasound catheter (Fig. 13).

## 9. HYPERTROPHIC OBSTRUCTIVE CARDIOMYOPATHY[8]

Symptomatic patients with HOCM are initially treated conservatively. When the symptoms remain despite medical therapy there are options like septal myectomy (surgical) or alcohol septal ablation (percutaneous). To

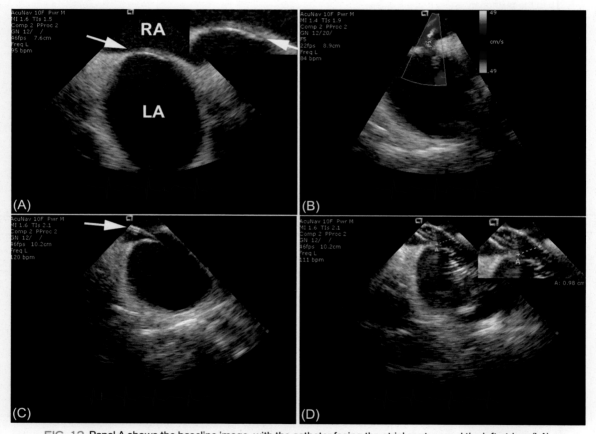

FIG. 12 Panel A shows the baseline image, with the catheter facing the atrial septum and the left atrium (LA). The *arrowhead* indicates the two leafs of the PFO blades with a small channel (narrow dark line between the leafs). Panel B: Color Doppler shows a clear "flaming" jet between the left and the right atrium, indicating leakage through the PFO. Panel C: The *arrowhead* indicates a guidewire passing through the PFO channel. Panel D: A sizing balloon is introduced over the guidewire and positioned in the PFO. After inflation the size appeared to be 9.8 mm.

*(Continued)*

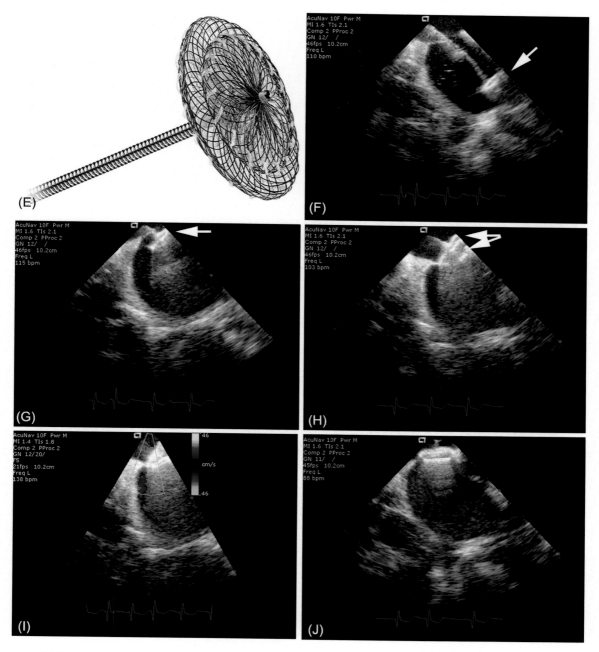

**FIG. 12, CONT'D** Panel E: Amplatzer PFO Occluder, Abbott Medical. For this patient, a 25-mm occluder was selected. Panel F: The occluder was introduced via a guiding catheter and partly deployed *(arrowhead)*. Panel G: The deployed disc was pulled back against the atrial septum *(arrowhead)*. Panel H: The second disc of the occluder was deployed in the right atrium *(arrowheads)*. Panel I: With both discs deployed, a final color Doppler test was performed prior to the final release. The smooth *blue color* represents the end-systolic flow in both the right and the left atrium. No jets were visible, compared with panel B. Panel J: Final result after releasing the occluder.

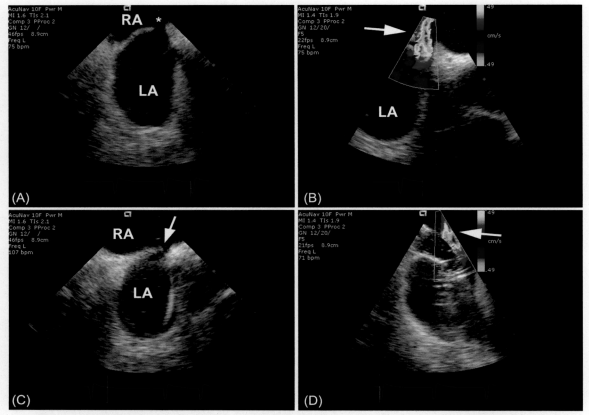

**FIG. 13** Panel A shows the baseline image, with the catheter facing the atrial septum and the left atrium (LA). The disruption of the atrial septum is visible near *. Panel B: Color Doppler shows the jet between the left and the right atrium, caused by the defect. Panel C: The *arrowhead* indicates a guidewire passing the ASD. Panel D: A sizing balloon is introduced over the guidewire and positioned in the ASD. The color jet *(arrowhead)* indicates leakage and thus incomplete inflation of the balloon.

*(Continued)*

guide an alcohol septal ablation procedure, the intracardiac ultrasound catheter is positioned in the right ventricle to visualize the ventricle septum and the left ventricle (Fig. 14).

## 10. LEFT ATRIAL APPENDAGE CLOSURE[9]
See Fig. 15.

## 11. 3D/4D DEVELOPMENTS (PHASED ARRAY ONLY)
Real-time 3D/4D imaging became possible when a sector width of 22 degrees was implemented to the phased array catheter (Siemens Acunav V). This paragraph shows miscellaneous examples from the author's archives. The positioning technique of the intracardiac catheter is not different compared to 2D imaging.

## 12. TRANSSEPTAL PUNCTURE
See Figs. 16 and 17.

## 13. CRYOABLATION; POSITIONING OF THE CRYOBALLOON
See Figs. 18–20.

## 14. PFO CLOSURE
See Fig. 21.

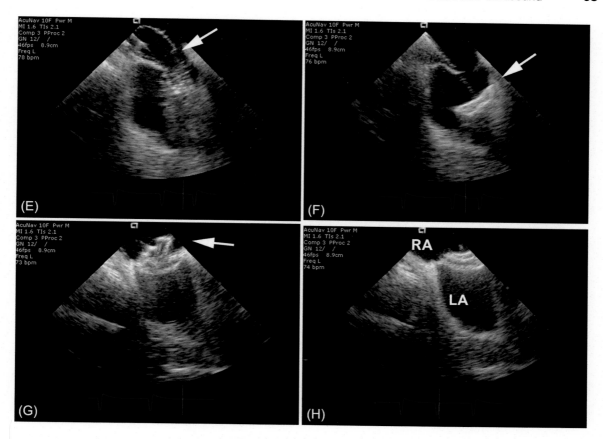

**FIG. 13, CONT'D**    Panel E: The sizing balloon is fully deployed and the required size of the occluder can be calculated. Panel F: The occluder was introduced via a guiding catheter and partly deployed *(arrowhead)*. Panel G: After deployment of the second disc of the occluder the secure placement of the occluder is checked by slightly tugging on the wire ("Minnesota wiggle") *(arrowhead)*. Panel H: Final result after the occluder was released.

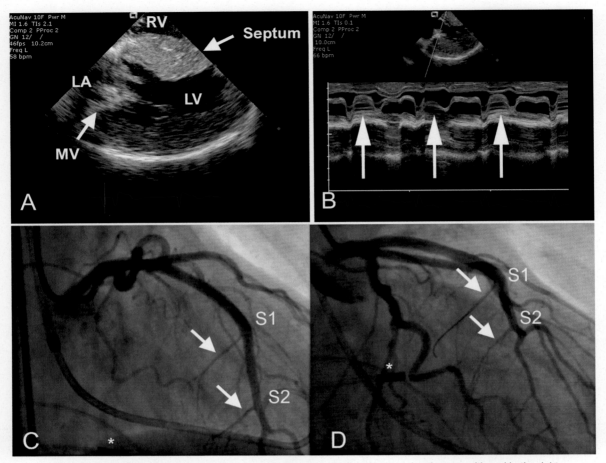

FIG. 14 Panel A: Baseline intracardiac ultrasound image, with the ultrasound catheter positioned in the right ventricle (RV) facing the thickened ventricle septum *(arrowhead)* and the left ventricle (LV). The second *arrowhead* indicates the mitral valve (MV). The left atrial (LA) outflow track is visible on the left side of the image. Panel B: Pulse Doppler tracing showing a systolic adverse movement (SAM, *arrowheads*) of the mitral valve, a common finding with HOCM. Panel C: Angiogram of the left coronary artery with two septal branches visible (S1 and S2, *arrowheads*). * indicates the ultrasound catheter in the right ventricle. Panel D: Angiogram of the left coronary artery in a different projection compared to C, with a guidewire positioned in the first septal branch (S1). * indicates the ultrasound catheter in the right ventricle.

*(Continued)*

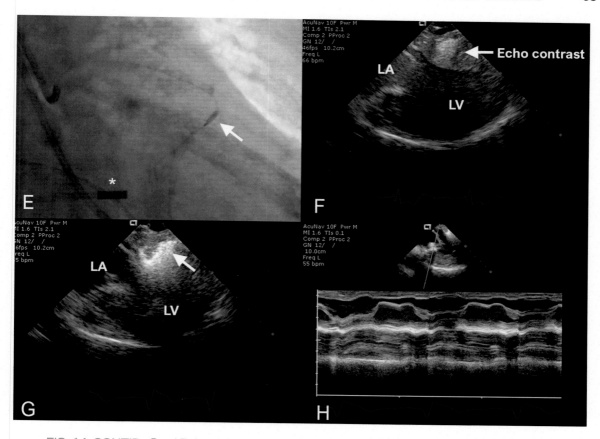

**FIG. 14, CONT'D**  Panel E: Angiogram showing a small over the-wire balloon inflated in the first septal branch *(arrowhead)*. While the balloon is inflated and the guidewire removed, echo contrast is injected prior to ablation through the balloon channel to check if the correct region in the septum will be affected. * indicates the ultrasound catheter in the RV. Panel F: Intracardiac ultrasound image after the administration of echo contrast. The bright shape in the ventricle septum *(arrowhead)* shows the region reached by the echo contrast. Panel G: Intracardiac ultrasound image after the administration of alcohol through the balloon channel. Due to the alcohol administration the microbubbles of the echo contrast increase in size, causing a stronger reflection on ultrasound *(arrowhead)*. Panel H: Pulse Doppler tracing showing immediate effect of the alcohol ablation; the SAM has disappeared (Compare panel B).

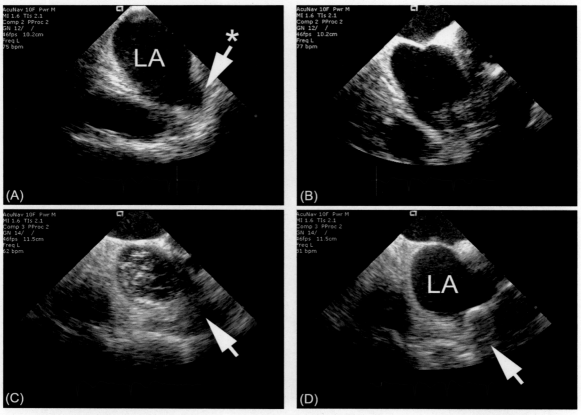

FIG. 15 Panel A: Baseline intracardiac ultrasound image. The *arrowhead* indicates the left atrial appendage (LAA). Panel B: Tenting (*) in the atrial septum during trans-septal puncture. Panel C: The PLAATO device is positioned and echo contrast is administered prior to release to assure that there is no leakage into the LAA. Panel D: Final result after PLAATO implantation. The *arrowhead* indicates the PLAATO device in situ. NB: PLAATO is discontinued and replaced by devices like Watchman (Boston Scientific).

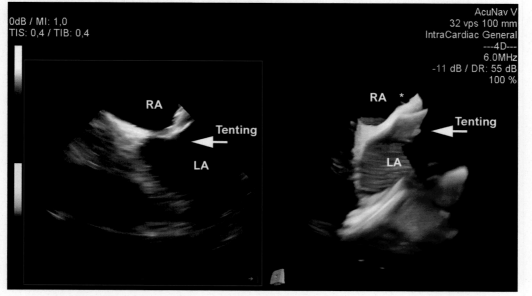

FIG. 16 The 2D image on the *left side* shows tenting of the needle in the atrial septum *(arrowhead)*. The image on the *right side* shows the real-time 3D image tilted so that the surface of the septum is visible from the side of the left atrium (LA). The *arrowhead* indicates bulging due to the tenting of the needle (*) from the right atrial (RA) side.

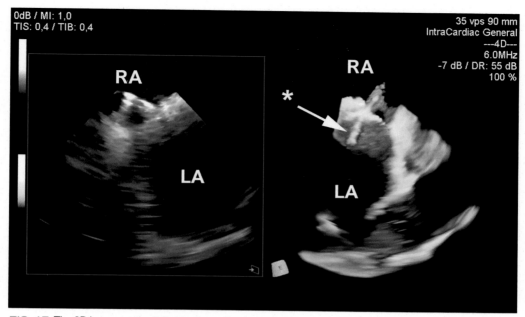

FIG. 17 The 2D image on the *left side* shows the catheter in the right atrium (RA). The image on the *right side* shows the real-time 3D image rotated and tilted so that the surface of the septum is visible from the side of the left atrium (LA). The *arrowhead* indicates the catheter (*) positioned through the fossa ovalis into the left atrium (LA).

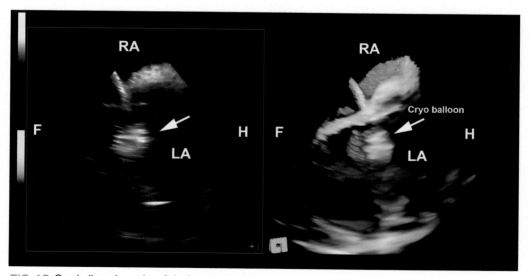

FIG. 18 Cryoballoon *(arrowhead)* deployed in the left atrium (LA). At the right-side is the real-time 3D image. *H* indicates the direction to the patient's head, F the direction to the patient's feet.

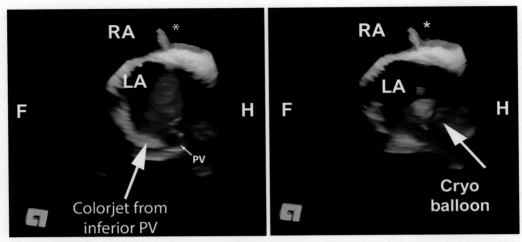

FIG. 19 Real-time 3D imaging with color Doppler to indicate the position of the inferior pulmonary vein (PV). The image on the *left side* in diastole shows the colorjet from the inferior PV *(arrowhead)*. The image on the *right side* in systole shows the conus shaped cryoballoon *(arrowhead)*. * indicates the catheter in the right atrium.

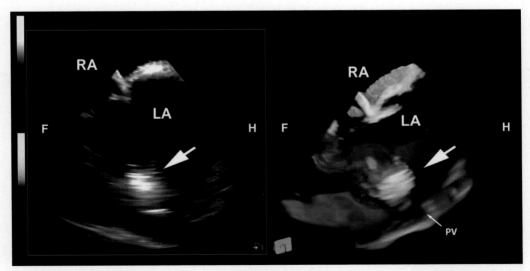

FIG. 20 2D *(left side)* and real-time 3D *(right side)* imaging with the cryoballoon *(arrowhead)* positioned at the inferior pulmonary vein (PV) ready for ablation. Note the visibility of the PV at the 3D image compared to 2D image.

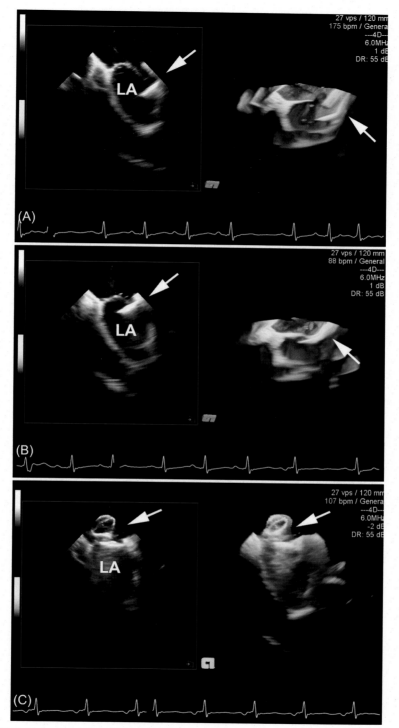

FIG. 21 2D *(left side)* and real-time 3D *(right side)* images of the final stages of a PFO closure with an Amplatzer PFO Occluder *(arrowhead)*. Panel A: The occluder with one disc deployed in the left atrium. Note the 3D "depth" effect of the occluder deep in the left atrium (LA). Panel B: Occluder is pulled against the septum. Panel C: "Minnesota wiggle" prior to permanent release of the occluder.

## REFERENCES

1. Seward JB, et al. Transvascular and intracardiac two-dimensional echocardiography. *Echocardiography*. 1990; 7 (4):457–464.
2. Saksena S, et al. A prospective comparison of cardiac imaging using intracardiac echocardiography with transesophageal echocardiography in patients with atrial fibrillation: the intracardiac echocardiography guided cardioversion helps interventional procedures study. *Circ Arrhythm Electrophysiol*. 2010; 3(6):571–577.
3. Vaina S, et al. Intracardiac echocardiography during interventional procedures. *EuroIntervention*. 2006; 1(4): 454–464.
4. Cismaru G, et al. Direct ICE imaging from inside the left atrial appendage during ablation of persistent atrial fibrillation. *Oxf Med Case Reports*. 2018; 2018(1):omx079.
5. Szili-Torok T, et al. Transseptal left heart catheterisation guided by intracardiac echocardiography. *Heart*. 2001; 86 (5):E11.
6. Hernandez F, et al. Intracardiac echocardiography and percutaneous closure of atrial septal defects in adults [Ecocardiografia intracardiaca en el cierre percutaneo de defectos del septo interauricular en adultos]. *Rev Esp Cardiol*. 2008; 61(5):465–470.
7. Rigatelli G, et al. Long-term outcomes and complications of intracardiac echocardiography-assisted patent foramen ovale closure in 1,000 consecutive patients. *J Interv Cardiol*. 2016; 29(5):530–538.
8. Pedone C, et al. Intracardiac echocardiography guidance during percutaneous transluminal septal myocardial ablation in patients with obstructive hypertrophic cardiomyopathy. *Int J Cardiovasc Interv*. 2005; 7(3):134–137.
9. Berti S, et al. Periprocedural intracardiac echocardiography for left atrial appendage closure: a dual-center experience. *JACC Cardiovasc Interv*. 2014; 7(9):1036–1044.

# CHAPTER 5

# Quantitative Virtual Histology for In Vivo Evaluation of Human Atherosclerosis— A Plaque Biomechanics-Based Novel Image Analysis Algorithm: Validation and Applications to Atherosclerosis Research ☆

PIOTR MUSIALEK[a] • WLADYSLAW DABROWSKI[a,b] • ADAM MAZUREK[a] • LUKASZ TEKIELI[a,b] • R. PAWEL BANYS[c,d] • JUAN RIGLA[e] • JUSTYNA STEFANIAK[f]

[a]Jagiellonian University, Department of Cardiac & Vascular Diseases, John Paul II Hospital, Krakow, Poland, [b]Department of Interventional Cardiology, John Paul II Hospital, Krakow, Poland, [c]Department of Radiology, John Paul II Hospital, Krakow, Poland, [d]Department of Physics and Applied Informatics, AGH University of Science and Technology, Krakow, Poland, [e]Department of Mathematics and Informatics, University of Barcelona, Barcelona, Spain, [f]Data Management and Statistical Analysis (DMSA), Krakow, Poland

## SUMMARY

Progress in biophysical modeling has recently identified several key plaque morphology features that play an important mechanistic role in plaque rupture, leading to often devastating clinical symptoms of the organ ischemia (stroke—in case of carotid artery atherosclerosis or myocardial infarction—in coronary atherosclerosis). Importantly, identification of these biophysics-relevant plaque features has been so far based on conventional (ex vivo) histological analyses while histological tissue processing affects both biological and biophysical characteristics of the atherosclerotic plaque components. Therefore, accurate (in situ), precise quantification of the key players that determine plaque biophysics and rupture risk is fundamental for increased precision of modeling in relation to better understanding the biologic processes and enhancing risk stratification.

Virtual histology intravascular ultrasound (VH-IVUS) enables high resolution (120 μm axial, 200–250 μm lateral for 20 MHz transducer; cf. rupture-prone fibrous cap thickness of <165–200 μm in the carotid plaque) and high-penetration (~5–10 mm) atherosclerotic plaque imaging. VH plaque components definition has been validated against human histology. Nevertheless, conventional VH image analysis and reporting methodology has several critical limitations; these are largely related to the qualitative (or semiquantitative) mode of current image evaluation algorithms. These fail to achieve acceptable reproducibility.

We have developed, for the first time, a fully quantitative VH (qVH) cross-sectional image analysis algorithm that is based on fundamental parameters of atherosclerotic plaque biomechanics that determine rupture risk. Cross-sectional plaque images were acquired with phased-array piezoelectric VH-IVUS transducers (Volcano-Philips).

---

☆Parts of the work outlined in the chapter have been presented to the European Society of Cardiology (ESC, Frontiers in Cardiovascular Biology) and the Transcatheter Cardiovascular Therapeutics (TCT) including the TCT Best of Vulnerable Plaque.

Intravascular Ultrasound. https://doi.org/10.1016/B978-0-12-818833-0.00005-9

QIvus software (v.2.0, Medis Medical Imaging Systems) was customized to measure the minimal fibrous cap (FC) thickness, peak confluent necrotic core (cNC) and peak confluent calcium (cCa) area, arc/angle, and thickness—beyond the conventional VH-IVUS parameters. To circumvent the VH-IVUS image acquisition limitations resulting from the patient heart rate and transducer pullback speed, the novel parameters were evaluated on a peak (i.e., maximal identified) basis; an approach that is routine in conventional histology assessment. The algorithm stability was validated by evaluating intertransducer and intratransducer (two transducers with two runs each, 21 consecutive plaques/patients; 52.4% symptomatic) and interobserver reproducibility of the key plaque component quantification.

Measurement variability of the novel parameters of cNC-area, cNC-thickness, cNC-arc, cCa-area, cCa-arc, and minFC thickness was low, respectively, 3.8/5.9%, 3.7/5.2%, 3.2/5.9%, 5.1/7.4%, 3.1/4.8%, and 4.7/3.2% (interobserver/intertransducer measurement). These variability levels were lower than those for conventional VH parameters. The between-observers and between-transducers measurement differences for the novel qVH IVUS parameters were not significant, consistent with a high stability of the novel algorithm.

In aggregate, we show that in vivo quantitative VH-IVUS evaluation of the carotid atherosclerotic plaque using novel parameters in relation to plaque biomechanics is feasible and is highly reproducible between the IVUS transducers, IVUS transducer runs, and between the analysts. These findings provide grounds for applying the novel qVH analysis algorithm—with its stable measurements of the minimal fibrous cap thickness and the peak confluent necrotic core and calcium area, thickness, and arc/angle—in cross-sectional and longitudinal studies of human atherosclerosis including modeling of plaque progression and clinical risk.

## LIST OF ABBREVIATIONS

| | |
|---|---|
| Ca | dense calcific plaque component |
| CAS | carotid artery stenting |
| cCa | confluent calcium |
| cNC | confluent necrotic core |
| CSA | (plaque) cross-sectional area ($EEM_{area} - lumen_{area}$) |
| EEM | external elastic membrane |
| FC | fibrous cap; min FC thickness—minimal fibrous cap thickness |
| FF | fibro-fatty plaque component |
| FT | fibrotic plaque component |
| ICA | internal carotid artery |
| NC | necrotic core |
| ROI | region of interest |
| TCFA | thin-cap fibroatheroma |

*Initial best estimation of the plaque components' contours, allowing subsequent iteration for elastic modulus assessment, is the basis for biophysical plaque stability determination.*

**GÉRARD FINET, JACQUES OHAYON AND COLLEAGUES[1,2]**

*The risk of plaque rupture depends on plaque type (composition) rather than plaque size (volume). The dynamic interplay between the actual plaque vulnerability and external stresses determines the particular moment and point of rupture.*

**ERLING FALK[3]**

*A novel computational framework [is now] able to integrate both morphological and biomechanical (fluid-dynamic and structural stresses) maps combining [the different crucial] factors (…) that lead to plaque rupture and cardiovascular events.*

**SIMONE BALOCCO AND COLLEAGUES[4]**

## 1. INTRODUCTION

Surprisingly, part of the medical community fails to understand that each symptomatic carotid artery atherosclerotic stenosis starts as an asymptomatic stenosis, and that "symptomatic" indicates, in most cases, that permanent cerebral ischemic damage has already occurred. As most strokes come without a clinical warning,[5] the guideline-indicated concept of "waiting for symptoms" (as a trigger for intervention) means for a majority of affected patients, waiting for a stroke, rather than maximazing preventive measures so that the stroke does not occur.

Atherosclerosis is a systemic inflammatory process but its serious, life-, or disability-threatening manifestations are focal.[6,7] Carotid atherosclerosis is a common pathology,[8] which is associated with ischemic stroke and has a profound impact on the life of individuals and their families.[9,10,28] One fundamental problem in understanding the relationship between the carotid atherosclerotic plaque and stroke is that the contemporary stroke risk stratification tools in carotid stenosis are nearly absent.[9,10,28] This is because the mechanisms of conversion from an asymptomatic to cerebral ischemia-causing lesion are poorly understood. As a result, for many carotid stenosis-related stroke victims, the treatment comes already too late to prevent disability.[9,11,12] The impact of stroke on the quality of life is so major that many stroke victims would prefer death rather than life after stroke.[11,13] Another problem is that even optimized pharmacologic management fails to offer effective protection against stroke risk[14–17] that, in vascular clinic patients on full medical treatment, currently reaches 2.0%–2.5% per year[14–16] and is cumulative over time (for instance, in a 50-year-old man, the 10-year stroke risk may reach 20% by the age of 60).[9,10] As the result, many of those who would benefit from revascularization (using surgery or neuroprotected carotid stenting) are not treated before the irreversible cerebral damage occurs[10] whereas for those with a large cerebral injury the intervention is often futile.[9,18] On a population basis, the intervention reduces stroke risk by about 50% over 10–15 years but it is not free of periprocedural risks.[19] Thus, patient and lesion risk stratification is crucial to identify those with maximal benefit in the context of the associated risk of the intervention.[9,20,21] Current risk stratification tools are based primarily on transcutaneous duplex ultrasound[28] that focuses mainly on the evaluation of stenosis severity, which is a very poor index of stroke risk[22,23] and provides only some very crude evaluation of the plaque.[21,24,28] Recent combined analysis from two large clinical trials in low-risk asymptomatic carotid stenosis showed a 5-year stroke risk of 7.8% with stenosis of 60%–69%, 7.4% with stenosis of 70%–79%, and 5.1% with stenosis of ≥80% by duplex ultrasound.[23] This indicates lack of discriminative power of the

stenosis severity once it exceeds ca. 50%. Moreover, angiographic evaluation of stenosis severity has fundamental reporting problems (note that 50% diameter stenosis is 75% area stenosis[25]) and its correlation with Doppler velocities through the lesion dependent on a range of variables[9] that may include the stenosis status of the contralateral carotid artery (absence/presence/severity)[26] or lesion length.[27] With this high significance of confounders and unclear definitions, the physician is expected to implement very different management pathways for lesions of "50% vs 60%" (undetermined how defined).[28] For plaque characterization, the transcutaneous duplex ultrasound visualization, similar to other commonly suggested noninvasive techniques such as computed tomography or magnetic resonance imaging has the fundamental limitation of poor image definition (resolution of 400–800 μm for typical transcutaneous ultrasound transducers, in-plane resolution ≈400–500 μm, and slice thickness ≈1000 μm for computed tomography and the magnetic resonance in-plane resolution of ≈300–600 μm with a slice thickness of ≈2000–3000 μm)[29–31] and artifacts such as blooming that occurs with calcifications.[32] As a result, these techniques perform best in mild stenoses or total occlusions.[33] Other carotid stenosis risk assessment tools such as the evaluation of microembolism risk with cerebral high-intensity transient signal monitoring on transcranial Doppler are impractical and have poorly defined and variable protocols, poor reproducibility, and lack of standardization.[10,34–36] All in all, the risk stratification based on a variety of current techniques remains fundamentally unsatisfactory for aiding clinical decision-making.[9,10,37–39]

A critical key to define plaque vulnerability[40–44] (i.e., the risk of rupture and the thrombotic event leading to adverse clinical events) is the accurate quantification of both the morphology and mechanical properties that affect the distribution of peak circumferential stress.[1,2,4] Plaque structural stress is determined by plaque composition, plaque architecture, and plaque geometry.[45] Recently, there has been fundamental progress in understanding the key players in plaque biomechanics that are associated with plaque vulnerability.[1–3,45–48] The precise size and circumference of rupture may be defined by where and when the heterogeneous distribution of tensile strength is overcome by the heterogeneous distribution of hemodynamic forces within the vessel.[3,49] The plaque biomechanics-related fundamental morphologic plaque components include the minimum fibrous cap thickness, and the area and angle of the necrotic core

(Fig. 1).[1,2,51–53] In addition, there is growing evidence that the amount and pattern of calcification is an important player in plaque biomechanics and the risk of rupture.[54–57] Initial best delineation of the plaque components' contours is fundamental, as it allows subsequent iteration for the assessment of elastic moduli and the peak circumferential stress locations.[1,2]

Identification of vulnerable plaques and defining their natural history remain major fundamental goals in contemporary cardiovascular medicine.[40–44,58] Intravascular imaging not only plays an important role in understanding the focal plaque processes leading to adverse ischemic events in symptomatic patients but also asymptomatic subjects at high risk of cardiovascular ischemic events may benefit from its use.[59]

Intravascular ultrasound with its axial resolution of 100–120 μm and tissue penetration of ≈5–10 mm in case of 20 MHz transducers (field of view ≈15 × 15 mm)[60–64] offers a direct in vivo assessment of the atherosclerotic plaque. IVUS has been enhanced with analysis of the backscattered signal frequency (rather than just amplitude), leading to ultrasonic identification of the four main components of plaque (fibrotic, fibro-fatty, calcific, and the necrotic core[65–68]). The virtual histology (VH) plaque component definition has been validated against conventional histology in humans, both for the coronary[68] and carotid[69] arteries. Recently, IVUS imaging enhanced by VH analysis has been shown to be predictive of coronary atherosclerotic clinical events.[70–72,113] Conventional VH analysis, however, suffers from suboptimal definitions associated with the current semiquantitative nature of image analysis.[65,66,73] Poor reproducibility of conventional VH assessment related to the inability to quantify the geographic distribution of key plaque components (on top of the dependence of the qualitative and semiquantitative analysis on the patient's heart rate and pullback speed[39,65,66]) has led, in longitudinal analyses, to false conclusions regarding supposed changes in the plaque structure that may represent, in fact, a change of the VH sampling spot ("slice").[39,74,75] In the coronary plaque evaluation, another significant VH limitation is related to the thickness of the coronary rupture-prone fibrous cap (≈65 μm on conventional histology imaging,[76] possibly corresponding to ≈100 μm in vivo[77]) that is below the IVUS axial resolution (≈120 μm for 20 MHz transducers). Importantly, this limitation does not apply to the carotid plaque imaging, where the critical (i.e., rupture-prone) fibrous cap thickness of 165–200 μm[78–80] exceeds the IVUS axial resolution by ≈60%–100%.[60,62,64]

In the present stage of the field, the quantitative analysis of VH frames has been limited to providing only a total (per cross-sectional area) amount of the plaque component(s) rather than the plaque biomechanics-relevant morphology features. Thus, a similar content of, for instance, the necrotic core component (confluent vs dissiminated) may correspond to entirely different plaque morphologies and plaque-associated clinical risk level (see Chapter 5 Section 4 and Fig. 7). As a consequence, the analytic tools available to-date could not show any relevant discriminative value.[81] This ignited research efforts to integrate the developments in understanding plaque biomechanics with the morphological distribution of the critical risk-associated plaque characteristics.[39,41,44,82,83]

We have developed and validated a novel VH image analysis algorithm that enables precise measurements of the plaque components that are vital to plaque biomechanics (Fig. 1), providing, for the first time, a reproducible stable diagnostic tool for in situ plaque assessment in the atherosclerotic adverse clinical event risk stratification.

FIG. 1 Schematic representation of key components (and their architecture in the atherosclerotic plaque) that are crucial to plaque biomechanics and rupture risk: minimal FC thickness, peak NC thickness, area and angle. According to the concept by Finet et al.[1] and Ohayon et al.[50]; modified; The necrotic core is shown in *red* consistent with its VH-IVUS tissue map representation whereas the fibrotic tissue is depicted in *green*.

## 2. METHODOLOGY

Based on fundamental parameters of the atherosclerotic plaque biomechanics that determine destabilization and rupture risk, we have developed a new quantitative VH (qVH) cross-sectional image analysis algorithm (Figs. 1 and 2). Cross-sectional plaque images were acquired with 20-MHz phased-array piezoelectric IVUS transducers (Volcano-Philips) consistent with general requirements and definitions in IVUS and VH-IVUS imaging.[62,65,77,84,85] The transducer pullback speed applied was deliberately slow (0.5 mm/s rather than 1.0 mm/s) to minimize the risk of missing the fundamental plaque morphology feature(s). Consistent with our pilot analysis, an additional run with the transducer deliberate stops at the sections of interest was routinely performed to maximize representative VH image captures that require ECG coupling and occur at the peak of the R wave.[39] It is important to note that VH-IVUS acquisition run performed with an automatic motorized pullback of the transducer at 0.5 mm/s translates, at the heart rate of 60 bpm and under an ideally stable movement of the transducer, into capturing a VH image "frame" every 0.5 mm longitudinally, i.e., with an interslice distance of 0.5 mm. This is fundamental in the context of, for instance, the "presence on three consecutive frames" requirement of lesion-type definition on conventional VH analysis[65] and it poses significant limitations of the volumetric analysis attempts using VH images.[86] Our sampling approach, with an additional run to ensure transducer stops in the segments of interest, mimics the conventional histology approach that is focused on the "peak" cross-sectional findings of interest (such as, e.g., the peak necrotic core size).[78-80,87,88] Moreover, this approach minimizes the risk of missing a crucial cross-sectional image due to a transducer "jump" at, for instance, bifurcation or a large eccentric plaque region.[89]

Image analysis was performed off-line. QIvus 2.0 software (Medis Medical Imaging Systems, the

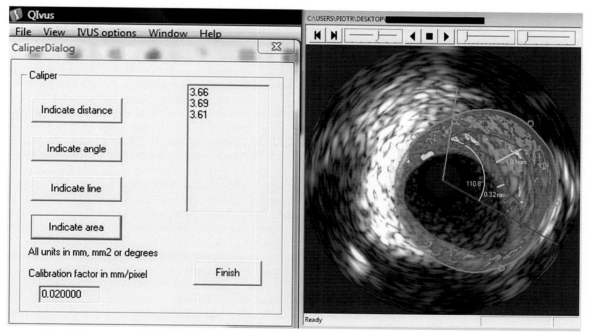

FIG. 2 A typical example of a carotid fibroatheroma VH-generated tissue map-based measurement of the key plaque-biophysics-related parameters as introduced in the novel analysis algorithm (qVH). After lumen delineation employing a simultaneous display of a matched flow color-coded image (cf. Fig. 3), each parameter is determined as an average of three measurements (QIvus 2.0 software customized interface on the *left* shows values for the peak NC area measurements). Image analysis performed with QIvus 2.0 software (Medis Medical Imaging Systems, the Netherlands) adapted to display the plaque biophysics-based parameters together with their numeric values.

Netherlands) was customized to enable measurements of the minimal fibrous cap (FC) thickness, peak confluent necrotic core (cNC), and peak confluent calcium (cCa) area, arc/angle, and thickness (typical example in Fig. 2)—beyond the conventional (automatic-provided) VH-IVUS parameters (i.e., the total content of NC, fibrofatty (FF), fibrotic (FT), and Ca component).

To enable precise lumen border definition in VH frames (that is critical for the fibrous cap thickness assessment), simultaneous matched image analysis of a VH slice and a corresponding ChromaFlo slice was routinely implemented (examples in Fig. 3). This approach, applied methodologically for the first time, overcomes one fundamental image processing limitation of conventional gray-scale and VH IVUS analysis that provides, on a single image, *either* colorized flow definition (ChromaFlo mode) *or* of the VH plaque components (VH mode).

The "peak content" VH frame selection was performed separately by each analyst, similar to the most representative slice selection on conventional histology by pathologists.[79,80,87,88,90,91] Three experienced analysts evaluated the images in a random order, with each IVUS run analyzed by two observers for the sake of reproducibility assessment. Each of the VH-IVUS novel parameters related to the plaque biomechanics and conventional (ex vivo) histology (min FC thickness, necrotic core peak area and angle, calcium peak area, and angle)[1,2,48,50,92,93] was measured, on a per-frame basis, three times by each the two independent analysts and the average values served the reproducibility evaluation (see Fig. 2 for a raw example). The algorithm reproducibility was tested in 21 consecutive plaques/patients (52.4% symptomatic), with two different transducers run in the evaluation of each plaque (Fig. 4). Data acquisition and evaluation were approved by our Institution's Ethics Committee, and all subjects provided an informed written consent.

Numeric data were expressed as absolute (distance, angle, area) and relative (per cross-sectional plaque area, CSA, proportionate content) values. Reproducibility of traditional and novel (qVH) measurements (inter- and intratransducer, inter- and intraobserver) was evaluated using the Bland-Altman method,[94] and was also expressed by providing absolute and relative measurement variability for all comparisons[95,96] (for raw images of the matched sections of interest see Fig. 4).

## 3. RESULTS

There were no adverse cerebral and no other clinical events associated with image acquisition.[97] The proposed qVH algorithm methodology was feasible for the evaluation of the biophysics-related plaque components in the sections of interest in all the 21 consecutive carotid plaques (21 consecutive patients) in the qVH algorithm reproducibility evaluation. The novel qVH algorithm feasibility was further confirmed in the analysis of over 200 plaques in a cross-sectional study in symptomatic and asymptomatic carotid atherosclerosis.[98] As given in the Chapter Section 2, the novel parameters were evaluated on a peak (i.e., maximal identified) basis to mimic conventional (ex vivo) histology assessment and circumvent the VH-IVUS image acquisition limitations resulting from the patient heart rate and transducer pullback speed (see the Chapter Section 4).

The novel algorithm reproducibility evaluation cohort showed cNC-area varying from 0.5 to $8.9\,mm^2$, cNC-thickness 0.4–2.1 mm, cNC-arc 25–241 degrees, cCa-area 0.3–$4.2\,mm^2$, and cCa-arc 44–268 degrees. Minimum FC thickness was 0.21–0.53 mm (see Table 1).

Bland-Altman analysis, performed between the different transducers and different observers, showed small measurement differences in the absence of any systematic measurement shift (see Fig. 5). Measurement relative differences are presented in Table 2 and Fig. 6.

Conventional measurements of the "total" per CSA content showed NC (i.e., total amount of "red"-depicted tissue per VH frame) in the range of 1.5–$12.5\,mm^2$ (7.1%–54.2%), FF 3.3–$28.2\,mm^2$ (17.1%–78.2%), FT 7.5–$33.2\,mm^2$ (47.3%–81.9%), and Ca 0.75–$4.9\,mm^2$ (1.2%–25.5%). Interobserver mean measurement variability of (conventional) absolute/proportionate peak plaque content was 4.2/7.2% for NC, 5.5/7.7% for FF, 5.2/8.2% for FT, and 5.0/6.2% for Ca whereas intertransducer mean measurement variability was, respectively, 7.2/10.2%, 7.2/8.4%, 6.1/7.0%, and 8.6/11.7%. Measurement variability was lower for the novel parameters of cNC-area, cNC-thickness, cNC-arc, cCa-area, cCa-arc, and min FC thickness, with mean interobserver/intertransducer measurement variability of 3.8/5.9%, 3.7/5.2%, 3.2/5.9%, 5.1/7.4%, 3.1/4.8%, and 4.7/3.2%, respectively.

For the novel algorithm individual parameters, the between-observers and between-transducers measurement differences were not statistically significant (see Figs. 5 and 6 and Tables 1 and 2).

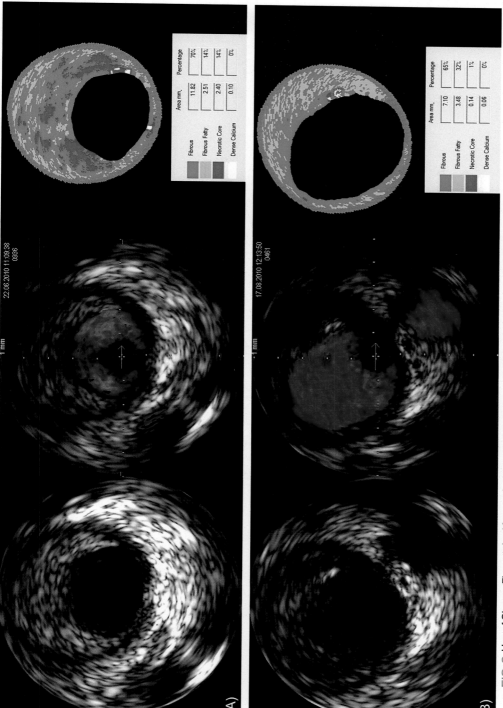

**FIG. 3** Use of ChromaFlo matched frames for increased precision of the lumen border detection for VH component analysis. Gray-scale IVUS *(left)* with simultaneous VH data acquisition, gray scale with flow coded in *red* (ChromaFlo, mid, requiring a separate IVUS run), and VH-IVUS visualization of the carotid atherosclerotic plaque content *(right)*. Examples illustrate the fundamental role of matched-frame analysis used by the present study to optimize the determination of the plaque luminal border. (A) and (B) Cross-sectional image of the carotid plaque exhibiting an optimal visualization of the external elastic membrane (EEM, seen as a dark rim in the gray-scale, left, and flow-coded, middle, image). Optimal visualization of EEM allows an accurate determination of the external border of the plaque, presented as a *light-gray* rim on the VH tissue map image *(right)*. Note that in (B) the luminal plaque interface is not seen on the gray-scale image. Adjunctive, simultaneous display of a paired flow-coded image *(red, middle panel)* allows precise delineation of the luminal interface of the plaque, and thus an accurate absolute and relative measurement of the VH-determined plaque components such as total NC, total FT, and total Ca *(right, inlet)*.

*(Continued)*

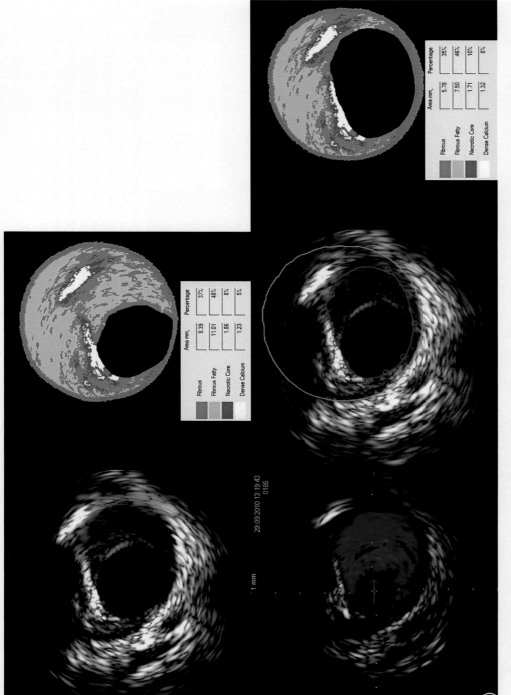

| | Area mm, | Percentage |
|---|---|---|
| Fibrous | 5.78 | 35% |
| Fibrous Fatty | 7.50 | 46% |
| Necrotic Core | 1.71 | 10% |
| Dense Calcium | 1.32 | 8% |

| | Area mm, | Percentage |
|---|---|---|
| Fibrous | 8.39 | 37% |
| Fibrous Fatty | 11.01 | 48% |
| Necrotic Core | 1.86 | 8% |
| Dense Calcium | 1.23 | 5% |

1 mm

29.09.2010 13:19:43
0185

(C)

FIG. 3, CONT'D   (B) and (C) illustrate the importance of using matched flow-coded images to avoid erroneous delineation of the luminal plaque border in constructing VH tissue maps (top panel) shows an example of erroneous delineation of the luminal plaque border occurring in the absence of guidance by a matched ChromaFlo frame, leading to a grossly wrong determination of the absolute and proportional content of VH plaque components). The luminal plaque interface becomes apparent when a paired flow-coded image is used as a guidance (bottom panel); part of the basis of the novel image analysis algorithm that we developed. Wrong determination of the plaque luminal border may have a major effect on VH determination of (absolute and relative) content of the respective plaque components (see inlets). Furthermore, failure to use flow-coding images as a guidance may introduce major errors in determining the position of confluent NC concerning the lumen and FC thickness. Note the necrotic core areas in the shoulders of plaque in (A); they can be quantified using the novel qVH algorithm but are not amenable to measurement when conventional VH analysis tools are used as they can provide only the total cross-sectional content (see text for comments).

**Transducer 1**

Run 1                Run 2

**Transducer 2**

Run 1                Run 2

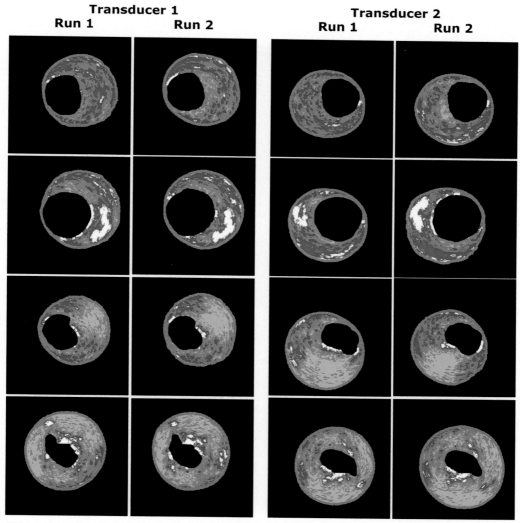

FIG. 4 Examples of raw data matching of the VH-determined plaque features (representative ROI images) specific for one carotid plaque and determined with two IVUS transducers (two pullbacks with each of the two transducers for each plaque, see Chapter Section 2). Because VH sampling is ECG coupled and occurs on the peak of R wave,[65, 66, 77] imaging (successful "hitting") vs missing some local plaque phenomena depends on the longitudinal position of the transducer in relation to the particular (local) plaque phenotype sampled at the peak of the R wave. Combined analysis of $\geq$1 IVUS run to characterize one lesion may (through an increased number of images) reduce the likelihood of missed visualization of certain plaque features, and may also limit apparent between-transducer differences that are due to relation of VH sampling to the transducer pullback speed and heart rate rather than "pure" image acquisition intertransducer differences (see Chapter Sections 2 and 4).

## 4. DISCUSSION

The qVH image analysis algorithm provides a novel fundamental tool for the evaluation of risk stratification based on plaque morphology. The algorithm enables precise measurements of the key players linked to plaque biomechanics that were demonstrated to determine the risk of rupture and symptomatic transformation. Thus key morphologic plaque parameters are now quantified, for the first time, in their biology and biophysics-related manner. This is very different from prior provisions and

**TABLE 1**
**VH-IVUS—Determined Carotid Plaque Compound Content in the Evaluated Lesions: Conventional "Total Content" Variables and Novel Quantitative (qVH) Parameters of Plaque Characteristics Relevant to Plaque Biophysics and Rupture Risk.**

| | TRANSDUCER 1 | | TRANSDUCER 2 | |
|---|---|---|---|---|
| | **Observer 1** | **Observer 2** | **Observer 1** | **Observer 2** |
| *Peak total content[a] per plaque CSA* | | | | |
| NC (mm$^2$) | 5.9 (1.5–11.0) | 5.8 (1.6–11.2) | 6.0 (1.4–12.5) | 5.9 (1.4–12.3) |
| NC (%) | 24.5 (7.1–54.2) | 25.6 (7.9–52.8) | 24.7 (7.8–52.0) | 24.3 (7.5–47.1) |
| FF (mm$^2$) | 10.7 (3.4–27.0) | 10.8 (3.3–28.2) | 10.8 (3.3–25.8) | 10.7 (3.4–26.4) |
| FF (%) | 38.4 (19.1–73.0) | 39.6 (17.1–78.2) | 40.0 (18.4–71.2) | 39.2 (19.7–74) |
| FT (mm$^2$) | 15.1 (8.6–32.4) | 15.0 (8.1–33.2) | 15.2 (7.5–30.1) | 15.3 (8.3–31.9) |
| FT (%) | 58.04 (48.2–70.9) | 59.82 (49.1–81.9) | 60.36 (50–71.1) | 58.14 (47–65.1) |
| Ca (mm$^2$) | 2.2 (0.9–4.2) | 2.2 (0.83–4.36) | 2.2 (0.77–4.9) | 2.2 (0.75–4.74) |
| Ca (%) | 7.6 (1.2–21.7) | 7.7 (1.2–23.2) | 7.6 (1.3–25.5) | 7.8 (1.3–24.1) |
| FF+NC (mm$^2$) | 16.7 (8.6–29.6) | 16.5 (8.7–30.2) | 16.9 (9.4–30.1) | 16.7 (9.6–28.2) |
| FF+NC (%) | 49.9 (35.2–76.1) | 49.4 (34.2–81.4) | 48.9 (35.8–77.0) | 49.5 (33.9–75.1) |
| *Confluent content[b]* | | | | |
| cNC area (mm$^2$) | 3.7 (0.5–8.7) | 3.7 (0.5–8.9) | 3.6 (0.5–8.5) | 3.6 (0.5–8.4) |
| cNC as % plaque CSA | 14.5 (1.8–37.6) | 14.7 (1.8–37.8) | 14.2 (1.8–37.4) | 14.5 (1.7–37.4) |
| cNC thickness (mm) | 1.24 (0.4–1.9) | 1.25 (0.4–2.1) | 1.22 (0.4–1.8) | 1.23 (0.5–1.9) |
| cNC arc (degrees) | 99.5 (28–235) | 100.3 (26–237) | 98.6 (25–240) | 98.3 (26–238) |
| cCa area(mm$^2$) | 1.5 (0.4–3.7) | 1.5 (0.3–3.6) | 1.5 (0.4–4.1) | 1.5 (0.4–4.2) |
| cCa arc (degrees) | 112.9 (44.4–268) | 112.8 (44.1–266) | 111.1 (45–258) | 110.8 (44.2–262) |
| *Minimum FC thickness[c] (mm)* | 0.32 (0.21–0.52) | 0.31 (0.22–0.51) | 0.31 (0.23–0.51) | 0.32 (0.21–0.53) |

Data are shown as mean (range) of peak value in 21 plaques (each measurement was performed thrice and the average value was used).
*CSA*, cross-sectional plaque area (CSA = EEM$_{area}$ − lumen$_{area}$); *NC*, necrotic core; *FF*, fibro-fatty component; *FT*, fibrotic component; *Ca*, calcium.
$P > .47$ in all analyzed parameters for the following comparisons:
    Transducer 1: Observer 1 vs Observer 1
    Transducer 2: Observer 1 vs Observer 2
    Observer 1: Transducer 1 vs Transducer 2
    Observer 2: Transducer 1 vs Transducer 2
[a]Conventional VH-IVUS parameters (evaluated as peak content per plaque).
[b]Peak confluent content per plaque.
[c]To be considered measurable, minFC thickness had to exceed the axial resolution of 20 MHz IVUS transducer (120 μm).

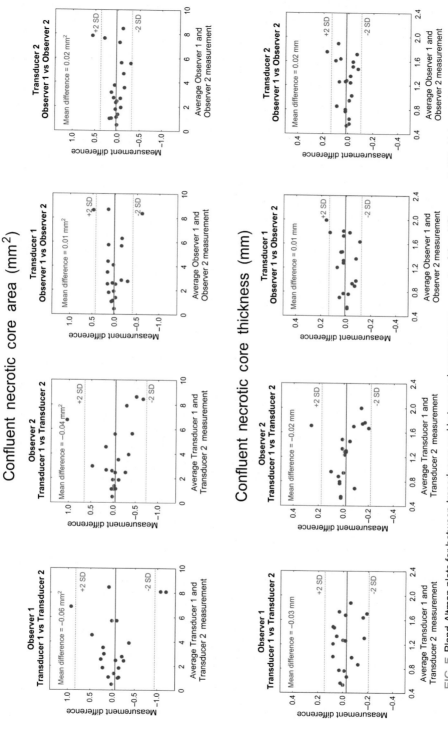

FIG. 5 Bland-Altman plots for between-transducers and between-observers measurements of the cNC-area and cNC-thickness. Note that the majority of data are within two standard deviations of the mean.

**TABLE 2**

**Absolute and Relative Measurement Variability for Conventional and Novel, Quantitative VH-IVUS Parameters of the Carotid Atherosclerotic Plaque Cross-Sectional Analysis.**

| | TRANSDUCER 1: OBSERVER 1 VS OBSERVER 2 | | | TRANSDUCER 2: OBSERVER 1 VS OBSERVER 2 | | | OBSERVER 1: TRANSDUCER 1 VS TRANSDUCER 2 | | | OBSERVER 2: TRANSDUCER 1 VS TRANSDUCER 2 | | |
|---|---|---|---|---|---|---|---|---|---|---|---|---|
| | abs | P | Relative | abs | P | Relative | abs | P | Relative | abs | P | Relative |
| *Conventional parameters[a]* | | | | | | | | | | | | |
| Total NC (mm$^2$) | 0.28 | .78 | **4.1%** (2.3–5.9) | 0.24 | .48 | **4.3%** (2.6–5.6) | 0.49 | .35 | **7.4%** (3.9–11) | 0.41 | .51 | **6.9%** (4.3–9.6) |
| Total NC (%) | 2.41 | .28 | **9.3%** (6.8–12) | 2.14 | .52 | **8.4%** (6.7–10) | 2.40 | .75 | **9.1%** (6.5–12) | 3.02 | .14 | **11.2%** (7.9–14) |
| Total FF (mm$^2$) | 0.59 | .34 | **5.1%** (3.9–6.3) | 0.63 | .33 | **5.8%** (4.1–7.5) | 0.83 | .56 | **7.3%** (4.6–9.5) | 0.78 | .38 | **7.1%** (5.3–8.9) |
| Total FF (%) | 3.20 | .22 | **8.1%** (6.7–9.5) | 2.84 | .33 | **7.2%** (5.4–8.9) | 3.41 | .24 | **7.9%** (4.2–12) | 3.70 | .76 | **8.8%** (5.9–11) |
| Total FT (mm$^2$) | 0.98 | .19 | **5.4%** (3.9–6.9) | 0.83 | .33 | **4.9%** (3.5–6.3) | 1.06 | .52 | **6.2%** (4.6–7.8) | 0.98 | .49 | **6.0%** (4.2–7.8) |
| Total FT (%) | 5.29 | .37 | **8.4%** (6.2–11) | 4.92 | .09 | **7.9%** (5.6–10) | 4.73 | .09 | **7.6%** (5.1–10) | 4.10 | .22 | **6.3%** (3.5–9.0) |
| Total Ca (mm$^2$) | 0.13 | .74 | **5.5%** (3.5–7.6) | 0.10 | .44 | **4.4%** (3.1–5.7) | 0.20 | .70 | **8.9%** (5.7–12) | 0.18 | .54 | **8.3%** (4.9–11) |
| Total Ca (%) | 0.71 | .34 | **6.9%** (4.7–9.1) | 0.56 | .49 | **7.5%** (5.4–9.7) | 1.05 | .79 | **11.9%** (7.9–15) | 0.88 | .62 | **11.4%** (8.4–14) |
| Total (FF+NC) (mm$^2$) | 0.82 | .65 | **5.6%** (4.0–7.1) | 0.71 | .78 | **4.7%** (3.4–5.9) | 1.02 | .84 | **6.4%** (5.0–7.8) | 0.97 | .44 | **6.6%** (4.3–8.9) |
| Total (FF+NC) (%) | 3.34 | .42 | **6.6%** (4.6–8.7) | 3.12 | .50 | **6.1%** (4.7–7.5) | 2.92 | .26 | **5.6%** (3.5–7.7) | 3.70 | .95 | **7.1%** (4.9–9.2) |
| *Confluent content* | | | | | | | | | | | | |
| cNC area (mm$^2$) | 0.15 | .12 | **3.9%** (2.7–5.0) | 0.12 | .54 | **3.7%** (2.0–5.4) | 0.29 | .57 | **6.6%** (5.6–8.5) | 0.23 | .57 | **5.2%** (3.3–7.0) |
| cNC area as (% CSA) | 0.93 | .80 | **6.5%** (4.1–8.8) | 0.91 | .21 | **7.2%** (5.3–9.0) | 0.81 | .19 | **5.9%** (3.4–8.5) | 1.06 | .72 | **8.2%** (5.1–11) |
| cNC thickness (mm) | 0.05 | .83 | **3.9%** (2.8–5.1) | 0.05 | .27 | **3.5%** (2.3–4.6) | 0.07 | .16 | **5.3%** (3.5–7.1) | 0.06 | .37 | **5.0%** (3.1–6.9) |
| cNC arc (degrees) | 2.63 | .65 | **3.1%** (2.3–3.8) | 3.32 | .74 | **3.3%** (2.2–4.5) | 5.98 | .59 | **6.3%** (4.8–7.9) | 4.47 | .39 | **5.5%** (3.6–7.3) |
| cCa area (mm$^2$) | 0.08 | .79 | **5.8%** (4.1–7.4) | 0.06 | .95 | **4.3%** (3.1–5.6) | 0.13 | .75 | **7.7%** (5.9–10) | 0.11 | .39 | **7.1%** (5.3–10) |
| cCa arc (degrees) | 2.79 | .88 | **2.9%** (1.9–3.9) | 2.99 | .87 | **3.3%** (1.9–4.8) | 5.75 | .24 | **5.3%** (3.3–7.2) | 4.59 | .17 | **4.3%** (2.3–6.3) |
| *min FC thickness (mm)* | 0.02 | .09 | **5.3%** (2.3–8.3) | 0.01 | .83 | **4.1%** (0.5–7.8) | 0.01 | .54 | **3.8%** (3.0–4.5) | 0.01 | .43 | **2.5%** (0.7–5.8) |

Absolute measurement variability (**abs**) is shown as mean whereas relative measurement variability is shown as mean (95% CI). Plaque component (*NC*, necrotic core; *FF*, fibro-fatty component; *FT*, fibrotic component; *Ca*, calcium; and *FF+NC*) peak content was taken as average of three measurements (see Chapter Section 2).

"*P*" values are given for absolute measurement variability between the observers (for a given IVUS transducer) or between the transducers (for a given observer), as indicated.

Absolute measurement variability was calculated as $|x_1 - x_2|$.

Relative measurement variability was calculated as $\dfrac{|x_1 - x_2|}{\max(x_1, x_2)} 100\%$.

[a]Total content per plaque CSA (cross-sectional plaque area; CSA=EEM$_{area}$ − lumen$_{area}$).

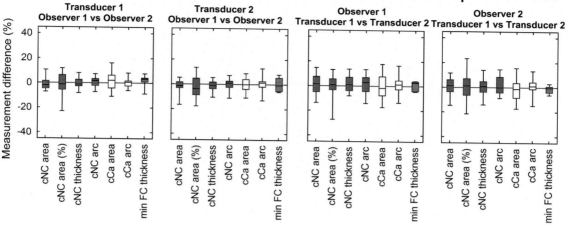

FIG. 6 Measurement differences for the qVH parameters: cNC (cNC-thickness, cNC-area, and cNC-arc), cCa (cCa-area and cCa-arc), and the minimal FC thickness. Box plots show median and 50% observations (box), with the whiskers representing the maximum and minimum value. Colors as in VH-IVUS tissue maps. Measurement (relative) difference was taken as $[(x_1 - x_2)/x_2] \cdot 100\%$. Image was analyzed with QIvus 2.0 software (Medis Medical Imaging Systems, the Netherlands) customized to display measurements of the new parameters with their numeric values.

assumptions of conventional (first generation) VH plaque content analyses. Those have relied on the values of a total (i.e., per slice) amount of any given VH plaque component that was presented as a blunt sum and was irrespective of any confluence, with little relevance to the risk of plaque rupture and thrombosis (cf. Fig. 3).

The VH-IVUS spectral tissue determination algorithm of the human atherosclerotic plaque, involving the four main components (F, FF, NC, and Ca) was trained from human histology, with which it was then shown to highly correlate.[68] An ex vivo validation study demonstrated an excellent accuracy between VH-IVUS and conventional histology in 184 plaques from 51 coronary arteries (fibrous component: accuracy 93.5%, sensitivity 95.7%, specificity 90.9%; FF: accuracy 94.1%, sensitivity 72.3%, specificity 97.9%; NC: accuracy 95.8%, sensitivity 91.7%, specificity 96.6%; dence calcium: accuracy 96.7%, sensitivity 86.5%, specificity 98.9%).[67,68] Furthermore, high correlations have been demonstrated in in vivo/ex vivo comparisons between human histology and VH-IVUS plaque composition from both carotid endarterectomies[99] and coronary atherectomy specimens.[100,101] In the carotid plaques, matching the VH-IVUS in vivo slices with conventional (ex vivo) histology focused of the "peak" findings of interest showed VH IVUS accuracy of 99.4% in thin-cap fibroatheroma, 96.1% for calcified thin-cap fibroatheroma, 85.9% in

fibroatheroma, 85.5% for fibrocalcific, 83.4% in pathological intimal thickening, and 72.4% for calcified fibroatheroma.[99] Nevertheless, at that time, no quantitative evaluation similar to our present approach (qVH) could be performed due to lack of a tool.

Several groups have recently attempted to employ conventional VH-IVUS imaging in carotid plaque risk stratification evaluation and to guide endovascular management. These endeavors, however, provided mostly inconsistent and/or impractical outcomes due to fundamental limitations of the traditional image processing and analytic approaches.[99,102–117]

The qVH algorithm, with its feasibility of use for the atherosclerotic plaque assessment in consecutive, unselected symptomatic and asymptomatic patients[96] and its high reproducibility between the transducers and observers (Figs. 5 and 6), provides a long-desired stable tool for an objective evaluation of the human atherosclerotic plaque in situ.[4,118–120] The algorithm overcomes some fundamental limitations of the prior simplistic approaches that have clearly suffered from poor reproducibility (starting with image sampling rate determination dependent on the patient's heart rate and pullback speed and poor border definitions), lack of quantitative analysis of the plaque key components in relation to their geographic distribution (see Fig. 7), and variable definitions of what constitutes, for instance, a thin-cap fibroatheroma.[39,65,66,77,121]

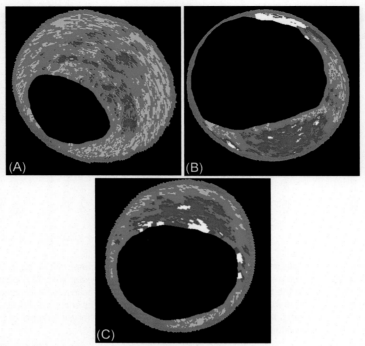

FIG. 7 Examples of plaques with a similar overall necrotic core (NC, coded in *red*) content but very different NC pattern and its relation to the fibrous cap morphology resulting in a very different pathophysiologic relevance. (A) diffuse necrotic core; (B) and (C) confluent necrotic core that plays an important role in plaque biomechanics and the risk of rupture. The fibrous cap is thick in (B) but absent by VH-IVUS imaging in (C), consistent with FC thickness below the VH-IVUS axial resolution of $\approx 120\,\mu m$ (in C— thin-cap fibroatheroma plaque morphology).

In conventional (phenotypic or semi-quantitative) VH assessment the variability of the VH "slice" sampling point has been typically mistaken for the change in plaque characteristics.[39,74,75] Subjective values of the purely qualitative or poor semiquantitative assessments have been resulting in suboptimal, and often clinically irrelevant, impractical measurements.[81,122,123] With the widespread rush for imaging of the human atherosclerotic plaque pathology in vivo,[75,124] any deeper understanding or acknowledgment of these fundamental problems of conventional VH image analysis tools has not been universal (albeit pointed out by some leading research groups[39,81,82,125]).

Until today, reproducibility of conventional measurements of the overall (i.e., per total cross-sectional plaque area rather than per the geographic site of interest cross-sectional image, see Fig. 7) content of the VH plaque components could be achieved only at the cost of ignoring the determination of the lumen-plaque interface and including the lumen to the wall for the sake of "a reproducible" image analysis,[126] resulting in pathologically and clinically irrelevant assessments.[127] This led to the long-term frustration, igniting efforts to develop quantitative VH analysis

algorithms.[82,128] These efforts, however, in contrast to the present completed approach, have only reached an early development stage of pixel value-based analysis attempts that relied on constructing, inconsistent with biology, "pixel neighbor rules" in attempts to determine "solid" component areas.[82]

The unique value of the qVH approach described in this chapter lies in several novel approaches that have been determined to be reproducible between the different transducers and observers. First, the image sampling in the regions of interest has been made independent of the patient's heart rate and the transducer pullback speed and it mimicks conventional histology assessment with one fundamental advantage of being performed in vivo. Because the VH-IVUS image ("frame/slice") acquisition is triggered by the R wave of the simultaneously recorded ECG, the frame frequency of VH-IVUS acquisition is dependent on the subject's heart rate. Thus, in the same patient (assuming a uniform pullback of the VH-IVUS imaging probe/catheter, see below) three frames recorded with a 0.5 mm/s automatic pullback would correspond to a longitudinal segment of interest (SOI) of 1.5 mm at the HR of 60 bpm whereas at the HR of

100 bpm SOI of 0.9 mm has already been represented by three VH-IVUS frames. Thus, if histological TCFA or CaTCFA is present in a longitudinal plaque segment of 1.3 mm, the plaque will be VH-IVUS classified as TCFA/CaTCFA if the patient's HR is 100 bpm (at least four VH-IVUS frames are then acquired in the SOI of 1.3 mm) but if the patient's HR is 60 bpm the same plaque will not be classified as TCFA/CaTCFA because only two VH-IVUS frames are captured within the same SOI of 1.3 mm. In such cases, the recognition (or lack of recognition) of the "vulnerable" plaque phenotype is HR dependent in addition to being dependent on the transducer pullback speed. At the HR of 60 bpm and pullback of 1 mm/s, the vulnerable plaque phenotype needs to be present in a longitudinal segment ≥3 mm to be represented by the three consecutive VH-IVUS frames to meet the definition requirement.[65,66]

There is no doubt that evaluation of the relative plaque composition requires accurate detection of the plaque external border and lumen border while the precise determination of the luminal border is a prerequisite for any necrotic core thickness measurement (see Fig. 3C).[125] Our first-time routine use of ChromaFlo-matched frames for VH lumen border determination enabled, in conjunction with customized quantitative analysis software, reproducible determination of the minimal fibrous cap thickness (Figs. 3 and 6, Table 2). While precise detection of the luminal plaque border is as important as EEM detection to enable quantification of the plaque components and their relative proportion, the precise detection of the luminal plaque border is particularly important for differentiating between the TCFA ("vulnerable") and fibroatheroma (FA, "stable" plaque) where the principal difference is the presence/absence of NC in direct contact with the lumen.[65,66,129] As intraluminal blood is coded as fibro-fatty or fibrous tissue by the VH spectral analysis software,[65,66] this becomes a false fibrous cap when included in the plaque on image processing.[130]

We have been able to introduce, for the first time, a routine and reproducible quantitative tool for the evaluation of the plaque features relevant to plaque biomechanics and conventional histology assessment. The qVH algorithm high-precision quantification of the atherosclerotic plaque key morphological players will now enable improved correlative analyses against new-generation noninvasive imaging modalities.[131–134]

The in situ plaque analysis using the novel qVH algorithm overcomes several fundamental limitations of conventional VH image analysis approaches that (i) are highly sensitive to the artifacts such as a (false) necrotic core determination in relation to calcifications [calcium-associated pseudo-necrotic core[135,136] (see Fig. 8)],

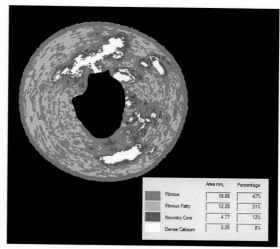

FIG. 8 Illustration of an erroneous impact that the calcium-associated pseudo-NC may have on conventional VH-IVUS analysis. The plaque NC is determined by a classic (automated) measurement of the total NC content on plaque CSA. VH-determined Ca areas usually have a rim of *red-colored* tissue—this is, largely, a misleading classification of the tissue that is artefactually codified as necrotic core by the current VH algorithm.[89, 135, 136]. In Fig. 8, most of the 12% NC arises from the calcium-associated "NC"; therefore, plaque analysis limited the use of automatically generated "NC total" content, which may lead to erroneous findings.

(ii) rely, for proportional assessment of the key elements (such as the necrotic core) content, on the highly variable (and sometimes impossible) determination of the external plaque border (Figs. 8–10),[122,123], and (iii) are sensitive to the patient heart rate and the catheter pullback speed,[39] and (iv) lack the ability to quantify the key plaque elements in relation to the plaque biomechanics to determine the risk of rupture.[139,140]

For measurements involving manual or semimanual delineation, the measurement variability is typically higher between the observers than within one observer.[141] In contrast, for qVH parameters, the interobserver measurement variability (2.9%–7.2%, Table 2) is not higher than that within any given observer (2.5%–7.7%, Table 2). Moreover, both are lower than peak variabilities seen with the conventional evaluation of the total cross-sectional content of VH plaque components (Table 2). This is consistent with the stability of the novel algorithm. This high level of intertransducer and interobserver agreement using qVH algorithm is critically important particularly as (which is not universally acknowledged), there is often a significant disagreement between pathologists when assessing conventional histology slices.[142–144]

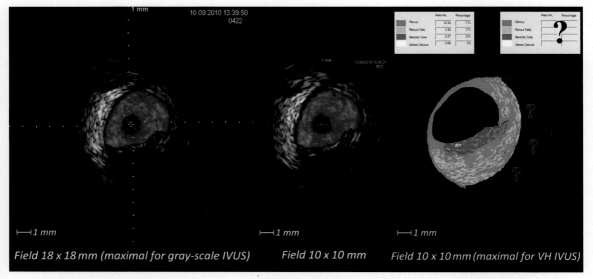

Field 18 x 18 mm (maximal for gray-scale IVUS)    Field 10 x 10 mm    Field 10 x 10 mm (maximal for VH IVUS)

FIG. 9 Illustration of the limitation in the conventional VH-IVUS plaque tissue map generation algorithm that is related to the size of VH-IVUS field diameter (currently 10 × 10 mm). Conventional gray scale or ChromaFlo imaging field is 18 × 18 mm in diameter (20 MHz EagleEye IVUS transducer, *left*). As the VH imaging field is limited to 10 × 10 mm in diameter, significant sections of a large, positively remodeled carotid plaque may be out of the VH field (*right*: the effect of limiting image field of 10 mm in diameter on the ChromaFlo image is shown in the middle). Construction of a tissue map for only part the plaque may lead to inaccurate determination of absolute and relative content of plaque components. This may affect not only quantitative but also qualitative classification of plaque phenotype (for instance, most definitions of thin-capped fibroatheroma, TCFA, require NC of at least 10% of the total plaque CSA[65, 77, 121, 137]—see also Fig. 10B). The images are on one scale.

Several limitations of the present work, resulting from the fundamental IVUS and VH limitations, need to be acknowledged. First, IVUS requires intravascular placement of the transducer.[59,145,146] Nevertheless, due to the physics-related limitations of noninvasive techniques (such as magnetic resonance imaging or computed tomography) resolution, noninvasive measurements of the critical plaque components cannot be presently performed with precision similar to the one we have obtained with our qVH algorithm that is dependent on intravascular imaging. Limitations inherent to VH spectral analysis include lack of thrombus determination algorithm,[125,130,147,148] lack of visualization of neovascularization and hemorrhage (whose roles, however, in "natural" atherosclerosis progression versus acute ischemic event triggering remain controversial),[47,48,149–151] and inability to evaluate the inflammatory process.[6,90,152–157] It should be noted that optical coherence tomography (OCT), an invasive technique that is sometimes claimed to constitute the intravascular imaging new gold standard, allows thrombus imaging and achieves image resolution greater than that of IVUS (12–15 μm

axial and 20–40 μm lateral vs 100–120 μm axial and 200–250 μm lateral). Nevertheless, due to its extremely poor tissue penetration (1.0–2.5 μm) OCT naturally fails in plaque imaging.[59,63,158] These OCT fundamental limitations might be partially overcome by future combined imaging incorporating both IVUS and OCT.[158,159] Finally, we deliberately resigned from longitudinal reconstructions because of the fundamental limitations of the VH signal sampling using automatic pullback that is affected by vessel pulsations and the transducer "jumps" over bifurcations or excentric plaque regions.[86,160] Furthermore, any longitudinal reconstructions are not part of conventional histology assessment and are of limited pathological relevance. This is because the plaque histology-related events occur mostly on the cross-sectional basis (cf., the concept of—and evidence for—the "culprit-of-the culprit" in acutely ischemic coronary lesions[161,162] or the relevance of critical cross-sectional findings in the carotid VICTORY study[163]) while the cross-sectional acute event consequences may be both cross-sectional (eg. lumen obstruction/occusion) and longitudinal (segment thrombosis with distal embolization).

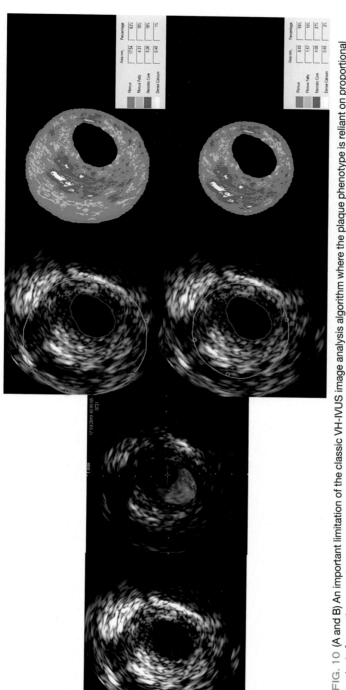

(A)

FIG. 10 (A and B) An important limitation of the classic VH-IVUS image analysis algorithm where the plaque phenotype is reliant on proportional content of, e.g., the necrotic core (note differences in the automatic determination of the proportional plaque contents in the inlets). These errors are due to the problematic external plaque border determination that may occur in the carotid artery. In contrast to the muscular arteries such as the coronary arteries (where the media appears routinely as an echolucent layer sandwiched between the more echodense intima and a hyperechoic adventitia and defines the outer limit of the atherosclerotic plaque[61, 138]), the internal carotid arteries show a marked variation in the muscular and elastic component[138] that may result in an indistinct interface between the media (EEM) and the adventitia.[85]

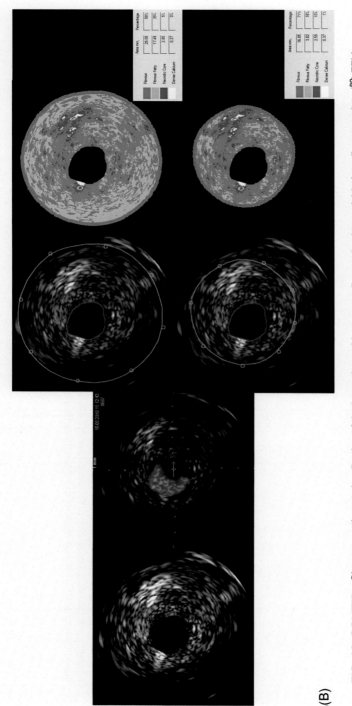

FIG. 10, CONT'D   Obscure or absent visualization of the internal carotid artery media may lead to variable (usually extrapolated[89]) EEM delineations, resulting in major differences in the determination of VH plaque total components due to the uncertain external plaque border location (compare the two inlets on the *right, top vs bottom*). Thus errors in border detection may lead to spurious conclusions about the plaque composition.[89, 125] Note differences in the automatic determination of the proportional plaque contents on the *right (top* inlet vs *bottom* inlet in A and B). The qVH algorithm, focused on the determination of the peak NC area, angle, and thickness, and minimal FC thickness is largely insensitive to the limitations illustrated in this figure (cf. Fig. 2).

(B)

Computational analyses of the plaque and vessel mechanics to date have suffered from lack of precise in vivo data to serve as a gold standard to validate the prediction models.[3,49,164–167] Moreover, with the unsatisfactory resolution of noninvasive imaging techniques and suboptimal balance between resolution and penetration in case of OCT, our understanding of the plaque biology and mechanics has been largely dependent, until today, on a "long-distance" inferring from conventional ex-vivo histology with its inherent errors and artifacts of tissue processing.[40,80,87,88,91,168,169] It is thus important to note that the in vivo quantitative plaque analysis introduced in this chapter enables plaque assessment that is free from the confounders and artifacts of histological tissue processing that may importantly change the biologic and mechanical tissue component relations as well as the particular content size (tissue shrinking, decalcification, shape changes and deformation, and content separations[170–173]). This opens a new era in our ability to understand the plaque biology in situ on the way toward highly desired personalized risk assessment.[52,119,174–179] This may be crucial for further development of the field. Particularly important is its impact on risk stratification because computational modeling of the plaque biomechanics requires, as a fundamental step, precise determination of the size and position of components that are then fed into the model.[1-3,47,50,180,181]

The simplicity of the novel qVH algorithm and its feasibility of use (Fig. 2) provides a tool for large-scale cross-sectional and longitudinal analyses of vast patient series, free from the fundamental limitations of several prior major endeavors such as PROSPECT, VIVA, VICTORY, ATHEROREMO, or IBIS-2.[70-72,163,182] Progress resulting from the algorithm application allows to overcome prior simplistic techniques that could demonstrate reproducibility only at the cost of, for instance, giving up any biologic relevance. Examples include introducing combined analyses of the plaque together with the lumen[126,127] or evaluation of the total cross-sections of VH component content rather than the qualitative analysis of the vital (confluent) content together with its geographical distribution.

The value of the novel algorithm has been recently demonstrated in depicting fundamental differences in the size of qVH plaque components in asymptomatic vs symptomatic plaques in a cross-sectional analysis of over 200 patients with carotid atherosclerosis in the CRACK-VH study[98] that showed feasibility to identify the asymptomatic plaques with symptomatic-like morphology. This important patient cohort is now the phase of the longitudinal follow-up.

Today there is a growing evidence that systemic (pharmacologic) management of atherosclerosis is not sufficient to prevent focal sudden ischemic events that include acute myocardial infarction, sudden death, and ischemic stroke.[9,14-17, 183] Nevertheless, it needs to be stressed that no focal vulnerable plaque management (such as surgical removal or endovascular isolation) should occur in the absence of the fundament of optimized systemic pharmacology.[9,18,20,28,184–186]

In conclusion, the novel qVH-IVUS algorithm enables, for the first time, quantitative in vivo evaluation of the atherosclerotic plaque components that play fundamental roles in plaque biology and biomechanics. The qVH plaque assessment is feasible for routine research use and is highly reproducible between the IVUS transducers, IVUS transducer runs, and between the analysts. The qVH analysis algorithm stability, with measurements of the minimal fibrous cap thickness and the peak confluent necrotic core area, thickness and arc/angle, and the peak calcium area, thickness, and arc/angle, is applicable in cross-sectional and longitudinal studies of human atherosclerosis. The algorithm provides a long desired, novel foundation for studying the atherosclerotic plaque biologic evolution in vivo with high precision,[120] extending knowledge needed for increased precision of computation modeling to predict plaque progression and vulnerability,[3,49,167] and guide clinical decision-making[20,52,178,185] ahead of the devastating clinical symptoms occurrence rather than in reaction to having crossed the point of irreversible injury at the clinical level.[12,17,179] A reproducible quantitative tool is now available for high-resolution in-situ evaluation of the atheroslerotic plaque features that is not only relevant to plaque biomechanics but is also free from the tissue processing artifacts of conventional (ex-vivo) histology.

## ACKNOWLEDGMENT

We are grateful to Professor Wieslawa Tracz and Professor Piotr Pieniazek for their continued encouragement and help in funding and conducting our research work.

## FUNDING

This work was supported by the Ministry of Science and Higher Education, Poland (National Committee for Scientific Research Project Grant N402184234), Polish Cardiac Society/Servier clinical research grant (Carotid Artery intravasculaR Ultrasound Study—CARUS), and

Polish Cardiac Society/ADAMED grant for basic research in atherosclerosis (Carotid arteRy plaque morphology And atherosClerosis biomarkers: Krakow Virtual Histology study—CRACK-VH) to PM.

# REFERENCES

1. Finet G, Ohayon J, Rioufol G. Biomechanical interaction between cap thickness, lipid core composition and blood pressure in vulnerable coronary plaque: impact on stability or instability. *Coron Artery Dis.* 2004; 15(1):13–20.
2. Ohayon J, Finet G, Le Floc'h S, et al. Biomechanics of atherosclerotic coronary plaque: site, stability and in vivo elasticity modeling. *Ann Biomed Eng.* 2014; 42(2):269–279.
3. Falk E. Why do plaques rupture? *Circulation.* 1992; 86(6 Suppl):30–42.
4. Balocco S, Gatta C, Alberti M, Carrillo X, Rigla J, Radeva P. Relation between plaque type, plaque thickness, blood shear stress, and plaque stress in coronary arteries assessed by X-ray angiography and intravascular ultrasound. *Med Phys.* 2012; 39(12):7430–7445.
5. Hackam DG, Kapral MK, Wang JT, Fang J, Hachinski V. Most stroke patients do not get a warning: a population-based cohort study. *Neurology.* 2009; 73:1074–1076.
6. Ross R. Atherosclerosis—an inflammatory disease. *N Engl J Med.* 1999; 340:115–126.
7. Libby P, Hansson GK. From focal lipid storage to systemic inflammation. *J Am Coll Cardiol.* 2019; 74(12):1594–1607.
8. De Weerd M, Greving JP, Hedblad B, et al. Prevalence of asymptomatic carotid artery stenosis in the general population: an individual participant data meta-analysis. *Stroke.* 2010; 41(6):1294–1297.
9. Musialek P, Hopf-Jensen S. Carotid artery revascularization for stroke prevention: a new era. *J Endovasc Ther.* 2017; 24(1):138–148.
10. Musialek P, Grunwald IQ. How asymptomatic is "asymptomatic" carotid stenosis? Resolving fundamental confusion(s)—and confusions yet to be resolved. *Pol Arch Intern Med.* 2017; 127(11):718–719.
11. Sturm JW, Donnan GA, Dewey HM, et al. Quality of life after stroke: the North East Melbourne Stroke Incidence Study (NEMESIS). *Stroke.* 2004; 35(10):2340–2345.
12. Trystula M, Tomaszewski T, Pachalska M. Health-related quality of life in ischaemic stroke survivors after carotid endarterectomy (CEA) and carotid artery stenting (CAS): confounder-controlled analysis. *Adv Interv Cardiol.* 2019; 15(2):226–233.
13. Cadilhac DA, Dewey HM, Vos T, Carter R, Thrift AG. The health loss from ischemic stroke and intracerebral hemorrhage: evidence from the North East Melbourne Stroke Incidence Study (NEMESIS). *Health Qual Life Outcomes.* 2010; 8:49.
14. Conrad MF, Boulom V, Mukhopadhyay S, et al. Progression of asymptomatic carotid stenosis despite optimal medical therapy. *J Vasc Surg.* 2013; 58(1):128–135.
15. Conrad MF, Michalczyk MJ, Opalacz A, Patel VI, LaMuraglia GM, Cambria RP. The natural history of asymptomatic severe carotid artery stenosis. *J Vasc Surg.* 2014; 60(5):1218–1226.
16. Kakkos SK, Nicolaides AN, Charalambous I, et al. Abbott AL for the asymptomatic carotid stenosis and risk of stroke (ACSRS) study group. *J Vasc Surg.* 2014; 59(4):956–967.
17. Paraskevas KI, Veith FJ, Ricco JB. Best medical treatment alone may not be adequate for all patients with asymptomatic carotid artery stenosis. *J Vasc Surg.* 2018; 68(2):572–575.
18. Musiałek P, Hopkins LN, Siddiqui AH. One swallow does not a summer make but many swallows do: accumulating clinical evidence for nearly-eliminated peri-procedural and 30-day complications with mesh-covered stents transforms the carotid revascularisation field. *Adv Interv Cardiol.* 2017; 13(2):95–106.
19. Halliday A, Harrison M, Hayter E, et al. 10-year stroke prevention after successful carotid endarterectomy for asymptomatic stenosis (ACST-1): a multicentre randomised trial. *Lancet.* 2010; 376:1074–1084.
20. Musialek P, Mazurek A, Trystula M, et al. Novel PARADIGM in carotid revascularisation: prospective evaluation of all-comer peRcutaneous cArotiD revascularisation in symptomatic and increased-risk asymptomatic carotid artery stenosis using CGuard™ MicroNet-covered embolic prevention stent system. *EuroIntervention.* 2016; 12(5):e658–e670.
21. Paraskevas KI, Ricco JB. The imperative need to identify stroke risk stratification models for patients with asymptomatic carotid artery stenosis. *J Vasc Surg.* 2018; 68(4):1277–1278.
22. Ahmed RM, Harris JP, Anderson CS, Makeham V, Halmagyi GM. Carotid endarterectomy for symptomatic, but "haemodynamically insignificant" carotid stenosis. *Eur J Vasc Endovasc Surg.* 2010; 40:475–482.
23. Derdeyn CP. Carotid stenting for asymptomatic carotid stenosis: trial it. *Stroke.* 2007; 38(2 Suppl):715–720.
24. Ricotta JJ, Aburahma A, Ascher E, Eskandari M, Faries P, Lal BK. Society for Vascular Surgery. Updated Society for Vascular Surgery guidelines for management of extracranial carotid disease. *J Vasc Surg.* 2011; 54(3):e1–31.
25. Ota H, Takase K, Rikimaru H, et al. Quantitative vascular measurements in arterial occlusive disease. *Radiographics.* 2005; 25(5):1141–1158.
26. Musialek P, Tekieli L, Pieniazek P, et al. Effect of contralateral occlusion on index internal carotid artery (ICA) duplex doppler velocities: quantitative analysis using intravascular ultrasound to match index ICA stenosis severity. *J Am Coll Cardiol.* 2012; 60(suppl B):B59.

27. Dabrowski W, Musialek P, Tekieli L, et al. Carotid lesion length effect on routine doppler velocities used to determine carotid stenosis cross-sectional severity: intravascular ultrasound-validated analysis in the CARUS study of 300 consecutive patients referred for potential carotid revascularization. *J Am Coll Cardiol*. 2013; 62(18 Suppl B):24.

28. Aboyans V, Ricco JB, Bartelink MEL, et al. 2017 ESC Guidelines on the Diagnosis and Treatment of Peripheral Arterial Diseases, in collaboration with the European Society for Vascular Surgery (ESVS): document covering atherosclerotic disease of extracranial carotid and vertebral, mesenteric, renal, upper and lower extremity arteries Endorsed by: the European Stroke Organization (ESO)The Task Force for the Diagnosis and Treatment of Peripheral Arterial Diseases of the European Society of Cardiology (ESC) and of the European Society for Vascular Surgery (ESVS). *Eur Heart J*. 2018; 39(9):763–816.

29. Devuyst G, Karapanayiotides T, Ruchat P, et al. Ultrasound measurement of the fibrous cap in symptomatic and asymptomatic atheromatous carotid plaques. *Circulation*. 2005; 111(21):2776–2782.

30. Dong L, Kerwin WS, Ferguson MS, et al. Cardiovascular magnetic resonance in carotid atherosclerotic disease. *J Cardiovasc Magn Reson*. 2009; 11:53.

31. U-King-Im JM, Young V, Gillard JH. Carotid-artery imaging in the diagnosis and management of patients at risk of stroke. *Lancet Neurol*. 2009; 8(6):569–580.

32. Thomas C, Korn A, Ketelsen D, et al. Automatic lumen segmentation in calcified plaques: dual-energy CT versus standard reconstructions in comparison with digital subtraction angiography. *AJR Am J Roentgenol*. 2010; 194(6):1590–1595.

33. Silvennoinen HM, Ikonen S, Soinne L, Railo M, Valanne L. CT angiographic analysis of carotid artery stenosis: comparison of manual assessment, semiautomatic vessel analysis, and digital subtraction angiography. *AJNR Am J Neuroradiol*. 2007; 28(1):97–103.

34. Abbott AL, Chambers BR, Stork JL, Levi CR, Bladin CF, Donnan GA. Embolic signals and prediction of ipsilateral stroke or transient ischemic attack in asymptomatic carotid stenosis: a multicenter prospective cohort study. *Stroke*. 2005; 36(6): 1128–1133.

35. Markus HS, King A, Shipley M, et al. Asymptomatic embolisation for prediction of stroke in the Asymptomatic Carotid Emboli Study (ACES): a prospective observational study. *Lancet Neurol*. 2010; 9(7):663–671.

36. Spence JD, Tamayo A, Lownie SP, Ng WP, Ferguson GG. Absence of microemboli on transcranial Doppler identifies low-risk patients with asymptomatic carotid stenosis. *Stroke*. 2005; 36(11):2373–2378.

37. Alsheikh-Ali AA, Kitsios GD, Balk EM, Lau J, Ip S. The vulnerable atherosclerotic plaque: scope of the literature. *Ann Intern Med*. 2010; 153:387–395.

38. Brinjikji W, Huston 3rd J, Rabinstein AA, Kim GM, Lerman A, Lanzino G. Contemporary carotid imaging: from degree of stenosis to plaque vulnerability. *J Neurosurg*. 2016; 124(1):27–42.

39. Musialek P. Virtual histology intravascular ultrasound evaluation of atherosclerotic carotid artery stenosis: time for fully quantitative image analysis. *J Endovasc Ther*. 2013; 20(4):589–594.

40. Finn AV, Nakano M, Narula J, Kolodgie FD, Virmani R. Concept of vulnerable/unstable plaque. *Arterioscler Thromb Vasc Biol*. 2010; 30(7):1282–1292.

41. Narula J, Nakano M, Virmani R, et al. Histopathologic characteristics of atherosclerotic coronary disease and implications of the findings for the invasive and noninvasive detection of vulnerable plaques. *J Am Coll Cardiol*. 2013; 61(10):1041–1051.

42. Sakakura K, Nakano M, Otsuka F, Ladich E, Kolodgie FD, Virmani R. Pathophysiology of atherosclerosis plaque progression. *Heart Lung Circ*. 2013; 22(6):399–411.

43. Stefanadis C, Antoniou CK, Tsiachris D, Pietri P. Coronary atherosclerotic vulnerable plaque: current perspectives. *J Am Heart Assoc*. 2017; 6(3):e005543.

44. Tarkin JM, Dweck MR, Evans NR, et al. Imaging atherosclerosis. *Circ Res*. 2016; 118(4):750–769.

45. Costopoulos C, Huang Y, Brown AJ, et al. Plaque rupture in coronary atherosclerosis is associated with increased plaque structural stress. *JACC Cardiovasc Imaging*. 2017; 10(12):1472–1483.

46. Makris GC, Nicolaides AN, Xu XY, Geroulakos G. Introduction to the biomechanics of carotid plaque pathogenesis and rupture: review of the clinical evidence. *Br J Radiol*. 2010; 83(993):729–735.

47. Teng Z, Brown AJ, Calvert PA, et al. Coronary plaque structural stress is associated with plaque composition and subtype and higher in acute coronary syndrome: the BEACON I (Biomechanical Evaluation of Atheromatous Coronary Arteries) study. *Circ Cardiovasc Imaging*. 2014; 7(3):461–470.

48. Teng Z, Sadat U, Brown AJ, Gillard JH. Plaque hemorrhage in carotid artery disease: pathogenesis, clinical and biomechanical considerations. *J Biomech*. 2014; 47(4):847–858.

49. Thondapu V, Bourantas CV, Foin N, Jang IK, Serruys PW, Barlis P. Biomechanical stress in coronary atherosclerosis: emerging insights from computational modelling. *Eur Heart J*. 2017; 38(2):81–92.

50. Ohayon J, Finet G, Gharib AM, et al. Necrotic core thickness and positive arterial remodeling index: emergent biomechanical factors for evaluating the risk of plaque rupture. *Am J Physiol Heart Circ Physiol*. 2008; 295(2):H717–H727.

51. Cardoso L, Weinbaum S. Changing views of the biomechanics of vulnerable plaque rupture: a review. *Ann Biomed Eng*. 2014; 42(2):415–431.

52. Virani SS, Ballantyne CM. From plaque burden to plaque composition: toward personalized risk assessment. *JACC Cardiovasc Imaging*. 2017; 10(3):250–252.

53. Wong KK, Thavornpattanapong P, Cheung SC, Sun Z, Tu J. Effect of calcification on the mechanical stability of plaque based on a three-dimensional carotid bifurcation model. *BMC Cardiovasc Disord.* 2012; 12:7.

54. Hoshino T, Chow LA, Hsu JJ, et al. Mechanical stress analysis of a rigid inclusion in distensible material: a model of atherosclerotic calcification and plaque vulnerability. *Am J Physiol Heart Circ Physiol.* 2009; 297(2):H802–H810.

55. Maldonado N, Kelly-Arnold A, Vengrenyuk Y, et al. A mechanistic analysis of the role of microcalcifications in atherosclerotic plaque stability: potential implications for plaque rupture. *Am J Physiol Heart Circ Physiol.* 2012; 303(5):H619–H628.

56. Paprottka KJ, Saam D, Rübenthaler J, et al. Prevalence and distribution of calcified nodules in carotid arteries in correlation with clinical symptoms. *Radiol Med.* 2017; 122(6):449–457.

57. Pini R, Faggioli G, Fittipaldi S, et al. Relationship between calcification and vulnerability of the carotid plaques. *Ann Vasc Surg.* 2017; 44:336–342.

58. Fleg JL, Stone GW, Fayad ZA, et al. Detection of high-risk atherosclerotic plaque: report of the NHLBI Working Group on current status and future directions. *JACC Cardiovasc Imaging.* 2012; 5(9):941–955.

59. Di Mario C, Moreno PR. Invasive coronary imaging: any role in primary and secondary prevention? *Eur Heart J.* 2016; 37(24):1883–1890.

60. Engeler CE, Ritenour ER, Amplatz K. Axial and lateral resolution of rotational intravascular ultrasound: in vitro observations and diagnostic implications. *Cardiovasc Intervent Radiol.* 1995; 18(4):239–242.

61. Lee JT, White RA. Basics of intravascular ultrasound: an essential tool for the endovascular surgeon. *Semin Vasc Surg.* 2004; 17(2):110–118.

62. Mintz GS, Nissen SE, Anderson WD, et al. American College of Cardiology Clinical Expert Consensus Document on Standards for Acquisition, Measurement and Reporting of Intravascular Ultrasound Studies (IVUS). A report of the American College of Cardiology Task Force on Clinical Expert Consensus Documents. *J Am Coll Cardiol.* 2001; 37(5):1478–1492.

63. Regar E. Invasive imaging technologies: can we reconcile light and sound? *J Cardiovasc Med.* 2011; 12(8):562–570.

64. Saunamäki KI. Virtual histology and the hunt for the vulnerable plaque. *Eur Heart J.* 2006; 27(24):2914–2915.

65. García-García HM, Mintz GS, Lerman A, et al. Tissue characterisation using intravascular radiofrequency data analysis: recommendations for acquisition, analysis, interpretation and reporting. *EuroIntervention.* 2009; 5(2):177–189.

66. König A, Margolis MP, Virmani R, Holmes D, Klauss V. Technology insight: in vivo coronary plaque classification by intravascular ultrasonography radiofrequency analysis. *Nat Clin Pract Cardiovasc Med.* 2008; 5(4):219–229.

67. Nair A, Kuban BD, Tuzcu EM, Schoenhagen P, Nissen SE, Vince DG. Coronary plaque classification with intravascular ultrasound radiofrequency data analysis. *Circulation.* 2002; 106(17):2200–2206.

68. Nair A, Margolis MP, Kuban BD, Vince DG. Automated coronary plaque characterisation with intravascular ultrasound backscatter: ex vivo validation. *EuroIntervention.* 2007; 3(1):113–120.

69. Diethrich EB, Irshad K, Reid DB. Virtual histology and color flow intravascular ultrasound in peripheral interventions. *Semin Vasc Surg.* 2006; 19(3):155–162.

70. Calvert PA, Obaid DR, O'Sullivan M, et al. Association between IVUS findings and adverse outcomes in patients with coronary artery disease: the VIVA (VH-IVUS in Vulnerable Atherosclerosis) Study. *JACC Cardiovasc Imaging.* 2011; 4(8):894–901.

71. Cheng JM, Garcia-Garcia HM, de Boer SP, et al. In vivo detection of high-risk coronary plaques by radiofrequency intravascular ultrasound and cardiovascular outcome: results of the ATHEROREMO-IVUS study. *Eur Heart J.* 2014; 35(10):639–647.

72. Stone GW, Mintz GS. Letter by Stone and Mintz regarding article, "unreliable assessment of necrotic core by virtual histology intravascular ultrasound in porcine coronary artery disease" *Circ Cardiovasc Imaging.* 2010; 3(5):e4.

73. Sinclair H, Veerasamy M, Bourantas C, et al. The role of virtual histology intravascular ultrasound in the identification of coronary artery plaque vulnerability in acute coronary syndromes. *Cardiol Rev.* 2016; 24(6):303–309.

74. Deliargyris EN. Intravascular ultrasound virtual histology derived thin cap fibroatheroma: now you see it, now you don't.... *J Am Coll Cardiol.* 2010; 55(15):1598–1599.

75. Kubo T, Maehara A, Mintz GS, et al. The dynamic nature of coronary artery lesion morphology assessed by serial virtual histology intravascular ultrasound tissue characterization. *J Am Coll Cardiol.* 2010; 55(15):1590–1597.

76. Burke AP, Farb A, Malcom GT, et al. Coronary risk factors and plaque morphology in men with coronary disease who died suddenly. *N Engl J Med.* 1997; 1276–1282.

77. Maehara A, Cristea E, Mintz GS, et al. Definitions and methodology for the grayscale and radiofrequency intravascular ultrasound and coronary angiographic analyses. *J Am Coll Cardiol Img.* 2012; 5:S1–S9.

78. Mauriello A, Sangiorgi GM, Virmani R, et al. A pathobiologic link between risk factors profile and morphological markers of carotid instability. *Atherosclerosis.* 2010; 208(2):572–580.

79. Redgrave JN, Gallagher P, Lovett JK, Rothwell PM. Critical cap thickness and rupture in symptomatic carotid plaques: the Oxford plaque study. *Stroke.* 2008; 39:1722–1729.

80. Virmani R, Ladich ER, Burke AP, Kolodgie FD. Histopathology of carotid atherosclerotic disease. *Neurosurgery.* 2006; 59(5 Suppl 3):S219–S227.

81. Alfonso F, Hernando L. Intravascular ultrasound tissue characterization. I like the rainbow but...what's behind the colours? *Eur Heart J.* 2008; 29(14):1701–1703.

82. Papaioannou TG, Schizas D, Vavuranakis M, Katsarou O, Soulis D, Stefanadis C. Quantification of new structural features of coronary plaques by computational post-hoc analysis of virtual histology-intravascular ultrasound images. *Comput Methods Biomech Biomed Engin.* 2014; 17(6):643–651.

83. Baroldi G, Bigi R, Cortigiani L. Ultrasound imaging versus morphopathology in cardiovascular diseases. Coronary atherosclerotic plaque. *Cardiovasc Ultrasound.* 2004; 2:29.

84. Mintz GS. Imaging controls. In: Mintz GS, ed. *Cardiovascular Ultrasound.* Taylor&Francis; 2005:14–15. [Chapter 1].

85. Nishimura RA, Edwards WD, Warnes CA, et al. Intravascular ultrasound imaging: in vitro validation and pathologic correlation. *J Am Coll Cardiol.* 1990; 16: 145–154.

86. Xu ZS, Lee BK, Park DW, et al. Relation of plaque size to compositions as determined by an in vivo volumetric intravascular ultrasound radiofrequency analysis. *Int J Card Imaging.* 2010; 26(2):165–171.

87. Mauriello A, Servadei F, Sangiorgi G, et al. Asymptomatic carotid plaque rupture with unexpected thrombosis over a non-canonical vulnerable lesion. *Atherosclerosis.* 2011; 218(2):356–362.

88. Redgrave JN, Lovett JK, Gallagher PJ, Rothwell PM. Histological assessment of 526 symptomatic carotid plaques in relation to the nature and timing of ischemic symptoms: the Oxford plaque study. *Circulation.* 2006; 113(19):2320–2328.

89. Alfonso F, Hernando L, Dutary J. Virtual histology assessment of atheroma at coronary bifurcations: colours at the crossroads? *EuroIntervention.* 2010; 6(3):295–301.

90. Sangiorgi G, Roversi S, Biondi Zoccai G, et al. Sex-related differences in carotid plaque features and inflammation. *J Vasc Surg.* 2013; 57(2):338–344.

91. Spagnoli LG, Mauriello A, Sangiorgi G, et al. Extracranial thrombotically active carotid plaque as a risk factor for ischemic stroke. *JAMA.* 2004; 292(15):1845–1852.

92. Gao H, Long Q. Effects of varied lipid core volume and fibrous cap thickness on stress distribution in carotid arterial plaques. *J Biomech.* 2008; 41(14): 3053–3059.

93. Hao H, Iihara K, Ishibashi-Ueda H, Saito F, Hirota S. Correlation of thin fibrous cap possessing adipophilin-positive macrophages and intraplaque hemorrhage with high clinical risk for carotid endarterectomy. *J Neurosurg.* 2011; 114:1080–1087.

94. Bland JM, Altman DG. Statistical methods for assessing agreement between two methods of clinical measurement. *Lancet.* 1986; 1:307–310.

95. Lui K-J. Relative difference measures: paired-sample data. In: Lui K-J, ed. *Statistical Estimation of Epidemiological Risk.* Wiley & Sons Ltd.; 2004:47–64.

96. Musialek P, Dabrowski W, Tekieli L, et al. Reproducibility of in vivo virtual histology intravascular ultrasound evaluation of the carotid atherosclerotic plaque using novel quantitative parameters. *J Am Coll Cardiol.* 2012; 60(suppl B):B59.

97. Musialek P, Pieniazek P, Tracz W, et al. Safety of embolic protection device-assisted and unprotected intravascular ultrasound in evaluating carotid artery atherosclerotic lesions. *Med Sci Monit.* 2012; 18(2):MT7–18.

98. Musialek P, Tekieli L, Pieniazek P, et al. Novel Quantitative Virtual Histology Parameters in the Asymptomatic and Symptomatic Carotid Atherosclerotic Plaque Imaging: data from the First 222 Patients in CRACK-VH Study. *J Am Coll Cardiol.* 2012; 60(suppl B):B57.

99. Diethrich EB, Margolis PM, Reid DB, et al. Virtual histology intravascular ultrasound assessment of carotid artery disease: the Carotid Artery Plaque Virtual Histology Evaluation (CAPITAL) study. *J Endovasc Ther.* 2007; 14(5):676–686.

100. Funada R, Oikawa Y, Yajima J, et al. The potential of RF backscattered IVUS data and multidetector-row computed tomography images for tissue characterization of human coronary atherosclerotic plaques. *Int J Card Imaging.* 2009; 25(5):471–478.

101. Nasu K, Tsuchikane E, Katoh O, et al. Accuracy of in vivo coronary plaque morphology assessment: a validation study of in vivo virtual histology compared with in vitro histopathology. *J Am Coll Cardiol.* 2006; 47(12): 2405–2412.

102. González A, López-Rueda A, Gutiérrez I, et al. Carotid plaque characterization by virtual histology intravascular ultrasound related to the timing of carotid intervention. *J Endovasc Ther.* 2012; 19(6):764–773.

103. Hitchner E, Zayed MA, Lee G, Morrison D, Lane B, Zhou W. Intravascular ultrasound as a clinical adjunct for carotid plaque characterization. *J Vasc Surg.* 2014; 59(3):774–780.

104. Hitchner E, Zhou W. Utilization of intravascular ultrasound during carotid artery stenting. *Int J Angiol.* 2015; 24(3):185–188.

105. Hishikawa T, Iihara K, Ishibashi-Ueda H, Nagatsuka K, Yamada N, Miyamoto S. Virtual histology-intravascular ultrasound in assessment of carotid plaques: ex vivo study. *Neurosurgery.* 2009; 65(1):146–152.

106. Inglese L, Fantoni C, Sardana V. Can IVUS-virtual histology improve outcomes of percutaneous carotid treatment? *J Cardiovasc Surg.* 2009; 50:735–744.

107. Irshad K, Millar S, Velu R, Reid AW, Diethrich EB, Reid DB. Virtual histology intravascular ultrasound in carotid interventions. *J Endovasc Ther.* 2007; 14(2):198–207.

108. Lee W, Choi GJ, Cho SW. Numerical study to indicate the vulnerability of plaques using an idealized 2D plaque model based on plaque classification in the human coronary artery. *Med Biol Eng Comput.* 2017; 55(8): 1379–1387.

109. Matsumoto S, Nakahara I, Higashi T, et al. Fibro-fatty volume of culprit lesions in virtual histology intravascular ultrasound is associated with the amount of debris during carotid artery stenting. *Cerebrovasc Dis.* 2010; 29(5):468–475.

110. Matsuo Y, Takumi T, Mathew V, et al. Plaque characteristics and arterial remodeling in coronary and peripheral arterial systems. *Atherosclerosis*. 2012; 223(2):365–371.

111. Matter CM, Stuber M, Nahrendorf M. Imaging of the unstable plaque: how far have we got? *Eur Heart J*. 2009; 30:2566–2574.

112. Schiro BJ, Wholey MH. The expanding indications for virtual histology intravascular ultrasound for plaque analysis prior to carotid stenting. *J Cardiovasc Surg*. 2008; 49(6):729–736.

113. Stone GW, Maehara A, Lansky AJ, et al. A prospective natural-history study of coronary atherosclerosis. *N Engl J Med*. 2011; 364(3):226–235.

114. Tamakawa N, Sakai H, Nishimura Y. Evaluation of carotid artery plaque using IVUS virtual histology. *Interv Neuroradiol*. 2007; 13(Suppl 1):100–105.

115. Timaran CH, Rosero EB, Martinez AE, Ilarraza A, Modrall JG, Clagett GP. Atherosclerotic plaque composition assessed by virtual histology intravascular ultrasound and cerebral embolization after carotid stenting. *J Vasc Surg*. 2010; 52(5):1188–1194.

116. Tsurumi A, Tsurumi Y, Hososhima O, Matsubara N, Izumi T, Miyachi S. Virtual histology analysis of carotid atherosclerotic plaque: plaque composition at the minimum lumen site and of the entire carotid plaque. *J Neuroimaging*. 2013; 23(1):12–17.

117. Yamada K, Yoshimura S, Kawasaki M, et al. Prediction of silent ischemic lesions after carotid artery stenting using virtual histology intravascular ultrasound. *Cerebrovasc Dis*. 2011; 32(2):106–113.

118. Burke AP, Virmani R, Galis Z, Haudenschild CC, Muller JE. The pathologic basis for new atherosclerosis imaging techniques. *J Am Coll Cardiol*. 2003; 41:1874–1886.

119. Falk E, Sillesen H, Muntendam P, Fuster V. The high-risk plaque initiative: primary prevention of atherothrombotic events in the asymptomatic population. *Curr Atheroscler Rep*. 2011; 13:359–366.

120. Stone GW, Maehara A, Mintz GS. The reality of vulnerable plaque detection. *JACC Cardiovasc Imaging*. 2011; 4(8): 902–904.

121. Surmely JF, Nasu K, Fujita H, et al. Coronary plaque composition of culprit/target lesions according to the clinical presentation: a virtual histology intravascular ultrasound analysis. *Eur Heart J*. 2006; 27: 2939–2944.

122. Obaid DR, Calvert PA, McNab D, West NE, Bennett MR. Identification of coronary plaque sub-types using virtual histology intravascular ultrasound is affected by inter-observer variability and differences in plaque definitions. *Circ Cardiovasc Imaging*. 2012; 5(1):86–93.

123. Siewiorek GM, Loghmanpour NA, Winston BM, Wholey MH, Finol EA. Reproducibility of IVUS border detection for carotid atherosclerotic plaque assessment. *Med Eng Phys*. 2012; 34(6):702–708.

124. Hermus L, Tielliu IF, Wallis de Vries BM, van den Dungen JJ, Zeebregts CJ. Imaging the vulnerable carotid artery plaque. *Acta Chir Belg*. 2010; 110:159–164.

125. Frutkin AD, Mehta SK, McCrary JR, Marso SP. Limitations to the use of virtual histology-intravascular ultrasound to detect vulnerable plaque. *Eur Heart J*. 2007; 28(14): 1783–1784.

126. Shin ES, Garcia-Garcia HM, Serruys PW. A new method to measure necrotic core and calcium content in coronary plaques using intravascular ultrasound radiofrequency-based analysis. *Int J Card Imaging*. 2010; 26(4):387–396.

127. Shin ES, Garcia-Garcia HM, Sarno G, et al. Reproducibility of Shin's method for necrotic core and calcium content in atherosclerotic coronary lesions treated with bioresorbable everolimus-eluting vascular scaffolds using volumetric intravascular ultrasound radiofrequency-based analysis. *Int J Card Imaging*. 2012; 28(1):43–49.

128. Kaul S, Diamond GA. Improved prospects for IVUS in identifying vulnerable plaques? *J Am Coll Cardiol Img*. 2012; 5:S106–S110.

129. Rodriguez-Granillo GA, García-García HM, Mc Fadden EP, et al. In vivo intravascular ultrasound-derived thin-cap fibroatheroma detection using ultrasound radiofrequency data analysis. *J Am Coll Cardiol*. 2005; 46(11):2038–2042.

130. Nasu K, Tsuchikane E, Katoh O, et al. Impact of intramural thrombus in coronary arteries on the accuracy of tissue characterization by in vivo intravascular ultrasound radiofrequency data analysis. *Am J Cardiol*. 2008; 101:1079–1083.

131. Obaid DR, Calvert PA, Gopalan D, et al. Atherosclerotic plaque composition and classification identified by coronary computed tomography: assessment of computed tomography-generated plaque maps compared with virtual histology intravascular ultrasound and histology. *Circ Cardiovasc Imaging*. 2013; 6(5):655–664.

132. Saba L, Potters F, van der Lugt A, Mallarini G. Imaging of the fibrous cap in atherosclerotic carotid plaque. *Cardiovasc Intervent Radiol*. 2010; 33:681–689.

133. Shinohara Y, Sakamoto M, Kuya K, et al. Assessment of carotid plaque composition using fast-kV switching dual-energy CT with gemstone detector: comparison with extracorporeal and virtual histology-intravascular ultrasound. *Neuroradiology*. 2015; 57(9):889–895.

134. Sun J, Zhao XQ, Balu N, et al. Carotid plaque lipid content and fibrous cap status predict systemic CV outcomes: the MRI substudy in AIM-HIGH. *JACC Cardiovasc Imaging*. 2017; 10(3):241–249.

135. Murray SW, Palmer ND. What is behind the calcium? The relationship between calcium and necrotic core on virtual histology analyses. *Eur Heart J*. 2009; 30(1):125.

136. Sales FJ, Falcão BA, Falcão JL, et al. Evaluation of plaque composition by intravascular ultrasound "virtual histology": the impact of dense calcium on the measurement of necrotic tissue. *EuroIntervention*. 2010; 6(3):394–399.

137. Kröner ES, van Velzen JE, Boogers MJ, et al. Positive remodeling on coronary computed tomography as a

marker for plaque vulnerability on virtual histology intravascular ultrasound. *Am J Cardiol.* 2011; 107: 1725–1729.

138. Zacharatos H, Hassan AE, Qureshi AI. Intravascular ultrasound: principles and cerebrovascular applications. *Am J Neuroradiol.* 2010; 31(4):586–597.

139. Chiocchi M, Chiaravalloti A, Morosetti D, et al. Virtual histology-intravascular ultrasound as a diagnostic alternative for morphological characterization of carotid plaque: comparison with histology and high-resolution magnetic resonance findings. *J Cardiovasc Med.* 2019; 20(5):335–342.

140. Wang L, Wu Z, Yang C, et al. IVUS-based FSI models for human coronary plaque progression study: components, correlation and predictive analysis. *Ann Biomed Eng.* 2015; 43(1):107–121.

141. Salonen R, Haapanen A, Salonen JT. Measurement of intima-media thickness of common carotid arteries with high-resolution B-mode ultrasonography: inter- and intra-observer variability. *Ultrasound Med Biol.* 1991; 17(3):225–230.

142. Cramer SF. International variation in histologic grading is large and persistent feedback does not improve reproducibility. *Am J Surg Pathol.* 2004; 28(2):273–275.

143. Ioachim HL. On variability, standardization, and error in diagnostic pathology. *Am J Surg Pathol.* 2001; 25(8): 1101–1103.

144. Rosai J. Borderline epithelial lesions of the breast. *Am J Surg Pathol.* 1991; 15(3):209–221.

145. Falk E, Wilensky RL. Prediction of coronary events by intravascular imaging. *J Am Coll Cardiol Img.* 2012; 5: S38–S41.

146. Fujii K, Hao H, Ohyanagi M, Masuyama T. Intracoronary imaging for detecting vulnerable plaque. *Circ J.* 2013; 77(3):588–595.

147. Badimon L, Vilahur G. Thrombosis formation on atherosclerotic lesions and plaque rupture. *J Intern Med.* 2014; 276(6):618–632.

148. Thim T, Hagensen MK, Bentzon JF, Falk E. From vulnerable plaque to atherothrombosis. *J Intern Med.* 2008; 263:506–516.

149. Chistiakov DA, Orekhov AN, Bobryshev YV. Contribution of neovascularization and intraplaque haemorrhage to atherosclerotic plaque progression and instability. *Acta Physiol.* 2015; 213(3):539–553.

150. Michel JB, Delbosc S, Ho-Tin-Noé B, et al. From intraplaque haemorrhages to plaque vulnerability: biological consequences of intraplaque haemorrhages. *J Cardiovasc Med.* 2012; 13(10):628–634.

151. Milei J, Parodi JC, Ferreira M, Barrone A, Grana DR, Matturri L. Atherosclerotic plaque rupture and intraplaque hemorrhage do not correlate with symptoms in carotid artery stenosis. *J Vasc Surg.* 2003; 38:1241–1247.

152. Battes LC, Cheng JM, Oemrawsingh RM, et al. Circulating cytokines in relation to the extent and composition of coronary atherosclerosis: results from the ATHEROREMO-IVUS study. *Atherosclerosis.* 2014; 236(1): 18–24.

153. Campbell IC, Suever JD, Timmins LH, et al. Biomechanics and inflammation in atherosclerotic plaque erosion and plaque rupture: implications for cardiovascular events in women. *PLoS ONE.* 2014; 9(11):e111785.

154. Musialek P, Tracz W, Tekieli L, et al. Multimarker approach in discriminating patients with symptomatic and asymptomatic atherosclerotic carotid artery stenosis. *J Clin Neurol.* 2013; 9(3):165–175.

155. Naghavi M, Libby P, Falk E, et al. From vulnerable plaque to vulnerable patient: a call for new definitions and risk assessment strategies: Part I. *Circulation.* 2003; 108(14): 1664–1672.

156. Naghavi M, Libby P, Falk E, et al. From vulnerable plaque to vulnerable patient: a call for new definitions and risk assessment strategies: Part II. *Circulation.* 2003; 108(15): 1772–1778.

157. Philipp S, Böse D, Wijns W, et al. Do systemic risk factors impact invasive findings from virtual histology? Insights from the international virtual histology registry. *Eur Heart J.* 2010; 31(2):196–202.

158. Huisman J, Hartmann M, von Birgelen C. Ultrasound and light: friend or foe? On the role of intravascular ultrasound in the era of optical coherence tomography. *Int J Card Imaging.* 2011; 27(2):209–214.

159. Tian J, Ren X, Vergallo R, et al. Distinct morphological features of ruptured culprit plaque for acute coronary events compared to those with silent rupture and thin-cap fibroatheroma: a combined optical coherence tomography and intravascular ultrasound study. *J Am Coll Cardiol.* 2014; 63(21):2209–2216.

160. Huisman J, Egede R, Rdzanek A, et al. Between-centre reproducibility of volumetric intravascular ultrasound radiofrequency-based analyses in mild-to-moderate coronary atherosclerosis: an international multicentre study. *EuroIntervention.* 2010; 5:925–931.

161. Legutko J, Jakala J, Mintz GS, et al. Virtual histology-intravascular ultrasound assessment of lesion coverage after angiographically-guided stent implantation in patients with ST elevation myocardial infarction undergoing primary percutaneous coronary intervention. *Am J Cardiol.* 2012; 109(10):1405–1410.

162. Legutko J, Jakala J, Mintz GS, et al. Radiofrequency-intravascular ultrasound assessment of lesion coverage after angiography-guided emergent percutaneous coronary intervention in patients with non-ST elevation myocardial infarction. *Am J Cardiol.* 2013; 112(12): 1854–1859.

163. Sangiorgi G, Bedogni F, Sganzerla P, et al. The virtual histology in CaroTids observational RegistrY (VICTORY) study: a European prospective registry to assess the feasibility and safety of intravascular ultrasound and virtual histology during carotid interventions. *Int J Cardiol.* 2013; 168(3):2089–2093.

164. Brown AJ, Teng Z, Evans PC, Gillard JH, Samady H, Bennett MR. Role of biomechanical forces in the

natural history of coronary atherosclerosis. *Nat Rev Cardiol.* 2016; 13(4):210–220.

165. Celeng C, Takx RA, Ferencik M, Maurovich-Horvat P. Non-invasive and invasive imaging of vulnerable coronary plaque. *Trends Cardiovasc Med.* 2016; 26(6): 538–547.

166. Li ZY, Howarth SP, Tang T, Gillard JH. How critical is fibrous cap thickness to carotid plaque stability? A flow-plaque interaction model. *Stroke.* 2006; 37(5): 1195–1199.

167. Tang D, Kamm RD, Yang C, et al. Image-based modeling for better understanding and assessment of atherosclerotic plaque progression and vulnerability: data, modeling, validation, uncertainty and predictions. *J Biomech.* 2014; 47(4):834–846.

168. Lovett JK, Gallagher PJ, Rothwell PM. Reproducibility of histological assessment of carotid plaque: implications for studies of carotid imaging. *Cerebrovasc Dis.* 2004; 18:117–123.

169. Lovett JK, Redgrave JN, Rothwell PM. A critical appraisal of the performance, reporting, and interpretation of studies comparing carotid plaque imaging with histology. *Stroke.* 2005; 36:1091–1097.

170. Buscema M, Hieber SE, Schulz G, et al. Ex vivo evaluation of an atherosclerotic human coronary artery via histology and high-resolution hard X-ray tomography. *Sci Rep.* 2019; 9(1):14348.

171. Dalager-Pedersen S, Falk E, Ringgaard S, Kristensen IB, Pedersen EM. Effects of temperature and histopathologic preparation on the size and morphology of atherosclerotic carotid arteries as imaged by MRI. *J Magn Reson Imaging.* 1999; 10(5):876–885.

172. Nasiri M, Janoudi A, Vanderberg A, et al. Role of cholesterol crystals in atherosclerosis is unmasked by altering tissue preparation methods. *Microsc Res Tech.* 2015; 78(11):969–974.

173. Salmhofer W, Rieger E, Soyer HP, Smolle J, Kerl H. Influence of skin tension and formalin fixation on sonographic measurement of tumor thickness. *J Am Acad Dermatol.* 1996; 34:34–39.

174. Chang HJ, Lin FY, Lee SE, et al. Coronary atherosclerotic precursors of acute coronary syndromes. *J Am Coll Cardiol.* 2018; 71(22):2511–2522.

175. Dweck MR, Doris MK, Motwani M, et al. Imaging of coronary atherosclerosis—evolution towards new treatment strategies. *Nat Rev Cardiol.* 2016; 13(9): 533–548.

176. Marso SP, Frutkin AD, Mehta SK, et al. Intravascular ultrasound measures of coronary atherosclerosis are associated with the Framingham risk score: an analysis from a global IVUS registry. *EuroIntervention.* 2009; 5: 212–218.

177. Naghavi M, Falk E, Hecht HS, et al. SHAPE Task Force. From vulnerable plaque to vulnerable patient—Part III: executive summary of the Screening for Heart Attack Prevention and Education (SHAPE) Task Force report. *Am J Cardiol.* 2006; 98(2A):2H–15H.

178. Papaioannou TG, Kalantzis C, Katsianos E, Sanoudou D, Vavuranakis M, Tousoulis D. Personalized assessment of the coronary atherosclerotic arteries by intravascular ultrasound imaging: hunting the vulnerable plaque. *J Pers Med.* 2019; 9(1):E8.

179. Paraskevas KI, Mikhailidis DP, Veith FJ, Spence JD. Definition of best medical treatment in asymptomatic and symptomatic carotid artery stenosis. *Angiology.* 2016; 67(5):411–419.

180. Dolla WJ, House JA, Marso SP. Stratification of risk in thin cap fibroatheromas using peak plaque stress estimates from idealized finite element models. *Med Eng Phys.* 2012; 34(9):1330–1338.

181. Kok AM, Speelman L, Virmani R, van der Steen AF, Gijsen FJ, Wentzel JJ. Peak cap stress calculations in coronary atherosclerotic plaques with an incomplete necrotic core geometry. *Biomed Eng Online.* 2016; 15(1):48.

182. Serruys PW, Garcia-Garcia HM, Buszman P, et al. Effects of the direct lipoprotein-associated phospholipase A(2) inhibitor darapladib on human coronary atherosclerotic plaque. *Circulation.* 2008; 118:1172–1182.

183. Hafiane A. Vulnerable plaque, characteristics, detection, and potential therapies. *J Cardiovasc Dev Dis.* 2019; 6(3):E26.

184. Ardissino M. Vulnerable plaques and vulnerable patients: the bright and dark side of the moon. *J Am Coll Cardiol.* 2018; 72(17):2091.

185. Serruys PW, Garcia-Garcia HM, Regar E. From postmortem characterization to the in vivo detection of thin-capped fibroatheromas: the missing link toward percutaneous treatment. *J Am Coll Cardiol.* 2007; 50(10):950–952.

186. Musialek P, Roubin GS. Double-layer carotid stents: from the clinical need, through a stent-in-stent strategy, to effective plaque isolation… the journey toward safe carotid revascularization using the endovascular route. *J Endovasc Ther.* 2019; 26(4):572–577.

## CHAPTER 6

# A State-of-the-Art Intravascular Ultrasound Diagnostic System

YUSUKE SEKI • YUKI SAKAGUCHI • AKIRA IGUCHI
Terumo Corporation, Tokyo, Japan

## 1. INTRODUCTION
### 1.1. Overview

We will describe the intravascular ultrasound (IVUS) diagnostic system from the viewpoint of hardware in this chapter. IVUS has become an indispensable tool for intravascular intervention due to the remarkable advances made since its introduction in the early 1970s.[1–4] It provides both quantitative and qualitative information about intravascular tissue such as vessel dimensions and plaque characteristics whereas conventional catheter angiography provides only a 2D silhouette of the lumen based on fluoroscopic X-ray images.[5–10] IVUS is thus used to optimize stent implantation and to minimize stent-related problems.[11] According to the recent guidelines on percutaneous coronary intervention (PCI), the use of IVUS is reasonable for assessing left main lesions.[12,13] It is estimated that more than 85% of PCIs performed in Japan are now done using IVUS.

An IVUS diagnostic system consists of three main devices: an imaging catheter, a motor drive unit (MDU), and a console, as shown in Fig. 1. An imaging catheter is a single-use device, while an MDU and a console are reusable. It is also notable that both the imaging catheter and the MDU are used in the clean field of a cath-lab operating room, while the console is used in the unclean field. In order to maintain sterility in the clean field, the MDU needs to be covered with a disposable sterile cover during an operation.

An ultrasound transducer placed at the tip of the imaging catheter transmits an ultrasound pulse signal radially, typically at 10–60 MHz, inside a blood vessel. The higher the frequency, the better the image quality, while the lower the frequency, the better the imaging of deep tissue due to deeper penetration. The ultrasound wave is reflected at multiple boundaries between different tissue types. An ultrasound transducer "receives" the reflected signals (ultrasound echo) after transmission, i.e., it converts the mechanical displacement in a piezoelectric single crystal caused by the reflected signals into an electrical signal. The electrical signal is then amplified by a preamp followed by signal processing in the console.

The intensity of the received ultrasound signals correlates to the properties of the corresponding tissue because ultrasound waves are reflected at a boundary of tissues with different acoustic impedances. Therefore, hard tissue such as calcium plaque appears brighter than soft tissue in an ultrasound image. The locations of different tissue types are identified on the basis of the duration between transmission and reception of the ultrasound signals.

There are two principles involved in performing intravascular radial scanning, namely, mechanical and electronic phased array scanning. Fig. 2 shows a schematic representation of a mechanical IVUS imaging catheter (AltaView, Terumo Corporation). Typical specifications of the IVUS imaging catheter are listed in Table 1. With mechanical scanning, a single transducer rotates inside a tubular imaging window. The imaging window is flushed with heparinized saline prior to use to avoid air contamination, which can prevent ultrasound imaging due to acoustic decoupling. The rotation speed is several thousand revolutions per minute (rpm) to acquire image data at 30–90 frames per second (fps). The MDU plays important roles in the pullback of the imaging catheter as well as the rotation of the mechanical scanning transducer through a flexible coil shaft. An electronic phased array imaging catheter typically has 64 transducers for radial scanning. The differences between mechanical and electronic phased array scanning are discussed in Section 2.

*Intravascular Ultrasound.* https://doi.org/10.1016/B978-0-12-818833-0.00006-0

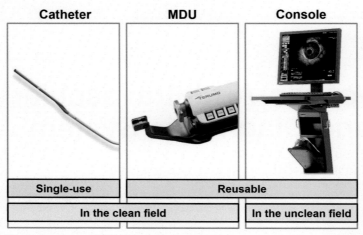

**FIG. 1** Three main components of IVUS diagnostic system. Imaging catheter (AltaView, Terumo Corporation) is a single-use device, while MDU and console (VISICUBE, Ueda Japan Radio Co., Ltd.) are reusable.

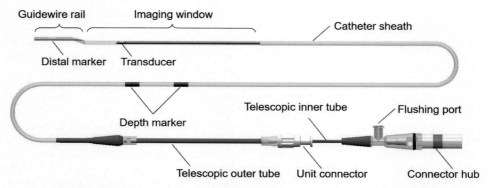

**FIG. 2** Schematic representation of mechanical IVUS imaging catheter with single guidewire rail (AltaView).

**TABLE 1**
Typical Specifications of IVUS Imaging Catheter (AltaView).

| Window diameter | Shaft diameter | Usable length | Frequency | Pullback speed | Rotation speed |
| --- | --- | --- | --- | --- | --- |
| 2.6 Fr (0.86 mm) | 3.0 Fr (1.01 mm) | 137 cm | 40–60 MHz | 0.5–9.0 mm/s | 1800–5400 rpm |

## 1.2. Procedure of IVUS Observation

This section briefly summarizes the IVUS observation procedure for a typical mechanical-type system. It is important to note that the procedure actually used should follow the instructions provided by device manufacturers.

### 1.2.1. Preparation

A package containing an IVUS imaging catheter typically includes an MDU cover, an extension tube with a three-

way stopcock, a flushing syringe, and a reservoir syringe as well as the imaging catheter housed in a tubular holder. After the console is turned on, the adapter attached to the MDU cover is connected to the MDU, which is gently covered with the sterile MDU cover.

The catheter holder is filled with heparinized saline to improve the lubricity of the catheter. The flushing syringe and reservoir syringe are connected to the extension tube via the three-way stopcock and then filled with heparinized saline. The extension tube is then connected to the flushing port of the catheter, and the telescopic inner tube is pulled out from the telescopic outer tube. Flushing is done by injecting heparinized saline to purge air from inside the catheter. This is important because air bubbles degrade image quality. After flushing, the catheter-side port on the three-way stopcock is closed.

The connector hub at the proximal end of the catheter is then mechanically and electrically connected to the MDU through the adapter of the MDU cover. An IVUS test image without sensor rotation is then checked to ensure that it shows concentric circles. The unit connector is then attached to the catheter clamp of the MDU through the MDU cover, and another IVUS test image with sensor rotation is checked. The telescopic inner tube is then pushed into the telescopic outer tube. The motor of the MDU is then turned off to stop sensor rotation. Finally, the catheter is carefully extracted from the holder.

### 1.2.2. Observation

The guidewire is lubricated using a cotton ball soaked in heparinized saline and then inserted along with a guiding catheter into the target blood vessel in the prescribed manner. Another flushing is generally done to eliminate air bubbles around the ultrasound transducer. The guidewire is passed through the guidewire rail of the imaging catheter, and the rail is inserted through the guiding catheter.

An IVUS test image with sensor rotation is then checked with the imaging catheter temporarily paused in the guiding catheter. Insertion of the imaging catheter is resumed after sensor rotation is stopped. The insertion distance is determined before fluoroscopy by using the depth markers on the catheter sheath (Fig. 2). The operator tracks the marker near the tip of the imaging catheter on a fluoroscopic X-ray image while pushing the catheter forward until it passes the target lesion.

Once the target lesion has been passed, the motor of the MDU is turned on again to resume sensor rotation, and an IVUS image is shown on the console. The operator can acquire multislice IVUS images while pulling the catheter back either manually or automatically using the MDU. Automatic pullback of the catheter by using the MDU enables the location and length of the target lesion to be identified. The pullback speed typically ranges from 0.5 to 9.0 mm/s. High-speed pullback reduces motion artifacts caused by heartbeat; it also shortens the procedure time.

### 1.2.3. Catheter removal

After the IVUS images have been observed, the telescopic inner tube is pushed back into the telescopic outer tube. The motor of the MDU is then turned off to stop sensor rotation, and the imaging catheter is carefully removed through the guiding catheter under fluoroscopic X-ray image guidance. It is returned to the catheter holder (refilled with heparinized saline) in preparation for the next observation.

At the end of the procedure, the imaging catheter is detached from the MDU. Finally, the MDU cover is removed from the MDU, and the console is turned off.

## 2. IVUS IMAGING CATHETER
### 2.1. Concept

An IVUS imaging catheter, i.e., a catheter-type ultrasound probe, is a device for directly observing the characteristics and morphology of blood vessels from within the vessel. The delivery of an IVUS catheter to a target lesion within a blood vessel requires certain functions and configurations. Moreover, the ultrasound transducer on the tip of the catheter needs to be able to visualize vascular formation. Therefore, the design concept in terms of both delivery and imaging is based on the functions listed in Table 2. In particular, trackability, pushability, and crossability are connected to each other.

### 2.2. Requirements for Catheter
### 2.2.1. Trackability

The trackability of a catheter is its ability to advance from an access site to a target lesion along a guidewire. An IVUS catheter is advanced through the target blood vessel along a guidewire, which runs through a guidewire rail. Therefore, the operator needs to pass the guidewire through the guidewire rail of the IVUS catheter before inserting it through a guiding catheter. The longer the guidewire rail, the better the trackability of the guidewire. The configuration of the guidewire rail needs to be carefully designed to minimize artifacts caused by the guidewire rail as well as the guidewire, which disturb ultrasound transmission.

An IVUS catheter typically advances through a guiding catheter from an access site to a coronary ostium. On the other hand, it advances from a coronary ostium to a target lesion through a coronary artery with complex vessel configuration, as shown in Fig. 3A, which varies from patient to patient.[14] Since the arterial configuration affords gentle curves from an access port in the

**TABLE 2**
Functions Required for IVUS Imaging Catheter With Regard to Delivery and Imaging.

| | Function | Element(s) | Needed |
|---|---|---|---|
| Delivery | Trackability | • Guidewire rail<br>• Catheter sheath<br>• Imaging core | To move back and forth along guidewire within guidewire rail in blood vessel |
| | Pushability | • Guidewire rail<br>• Catheter sheath<br>• Imaging core | To transmit force applied by hand to distal end |
| | Crossability | • Guidewire rail<br>• Catheter sheath<br>• Imaging core<br>• Coating | To smoothly pass through stenosis lesion |
| | Visibility | • Radiopaque markers<br>• Depth markers | To visualize catheter position under fluoroscopy |
| | Safety | • Imaging catheter | To safely deliver catheter within blood vessel |
| Imaging | Electrical signal transmission | • Imaging core | To transmit electrical signal between transducer and MDU |
| | Ultrasound signal transmission | • Transducer<br>• Imaging window | To transmit ultrasound signal between transducer and target |
| | Radial scan | • Coil shaft | To acquire cross-sectional images from transducer rotation |
| | Longitudinal scan | • Telescopic inner/outer tubes | To acquire longitudinal images from pullback |

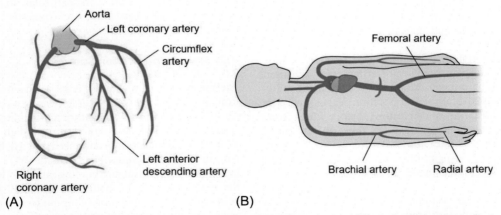

FIG. 3 Schematic representation of (A) coronary arterial configuration and (B) arteries from access site to coronary ostium.

radial, brachial, or femoral artery to the coronary ostium, except for the aortic arch, as shown in Fig. 3B, the IVUS catheter needs to be highly flexible only in the distal part to be able to deform along a coronary artery.

Flexibility in the distal part is necessary not only for a guidewire rail but also for a catheter sheath and an imaging core of an IVUS catheter. Therefore, a guidewire rail

and a catheter sheath consist of a flexible polymer material (e.g., polyethylene or nylon). By contrast, an imaging core consists of a metallic coil to acquire torque transmission, called torquability.

A typical example of a mechanical IVUS imaging catheter with a single guidewire rail is shown in Fig. 2. The rail is attached to the distal side of the imaging window in this case. With a single guidewire rail attached to the distal side of imaging window, the distance between the tip and the transducer affects the trackability. A long guidewire rail ahead of the transducer is better in terms of trackability, as mentioned above, but the tip needs to be advanced further past the target lesion.

### 2.2.2. Pushability

Pushability of a catheter is its ability to transmit force applied by an operator's hand to distal end. The usable section of a catheter is designed in view of the shape from the access port (the proximal end of the Y-connector attached to the guiding catheter) to the coronary lesion. For example, length and diameter of the usable section of AltaView (Fig. 2) are 1370 mm and 2.6–3.0 Fr, respectively (0.86–1.01 mm). A stiff material (e.g., super engineering plastic or stainless steel) is thus used for the proximal part of a catheter for enough pushability. An imaging core also contributes to enhance stiffness of an IVUS catheter. It is notable that stiffness of the distal part of an IVUS catheter is important to transmit force to distal end. Mechanically speaking, the distal end of the catheter needs to be flexible to improve trackability while the proximal end needs to be stiff to improve pushability. The design should thus optimize the balance between these conflicting properties.

The flexible part of the catheter needs to be longer than the longitudinal scanning distance from distal to proximal, or pullback length (typically 150 mm), since the imaging window is included in the flexible part. It also needs to be longer than the longest coronary artery. The junction between the flexible and rigid materials should not create an abrupt change in stiffness, which can create a kink in the catheter.

Pushability can be improved by making better use of a guidewire rail as a support of the flexible part of the IVUS catheter. Fig. 4 shows an IVUS imaging catheter with two guidewire rails (Navifocus WR, Terumo Corporation). In the double-rail configuration, the first guidewire rail is attached to the distal side of the imaging window, and the second one is attached to the proximal side. The double guidewire rail improves the pushability of the imaging catheter by leveraging support from the guidewire.

With a double guidewire rail, a short tip (short guidewire rail) is feasible because of the long guidewire rail proximal to the imaging window. This type of IVUS catheter is thus suitable for observing chronic total occlusion lesions, for which an IVUS catheter needs pushability as well as trackability.[15]

### 2.2.3. Crossability

Crossability of a catheter is its ability to smoothly pass through stenotic lesion. As mentioned above, both trackability and pushability affect crossability. On the other hand, friction between the catheter and the blood vessel wall is an ongoing problem in the field of catheter intervention. A lubricious coating made of hydrophilic polymer is thus typically applied to the outer surface of the catheter to reduce this friction.

It is obviously better to make outer diameter of a catheter as small as possible in terms of crossability. Tapered tip of an IVUS catheter is another effective design to improve crossability. For an IVUS catheter to be able to pass through a vascular stenosis, the difference between the outer diameter of the guidewire and the inner diameter of the guidewire rail must be optimized. The lumen size selected for the guidewire rail

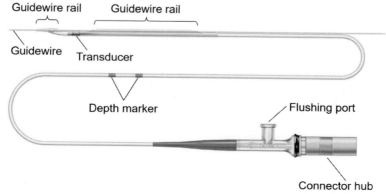

FIG. 4 Schematic representation of IVUS imaging catheter with two guidewire rails (Navifocus WR, Terumo Corporation). The use of two guidewire rails improves pushability of imaging catheter.

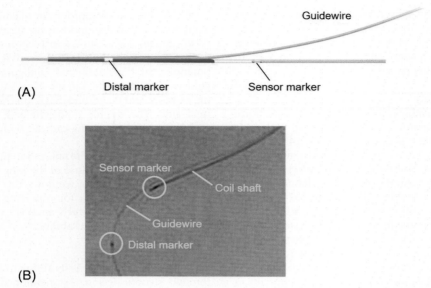

FIG. 5 (A) Radiopaque markers on IVUS catheter and (B) markers viewed under fluoroscopy.

should be in accordance with the outer diameter of the guidewire to be used, typically 0.014 in. in the case of PCI.

### 2.2.4. Visibility

An IVUS catheter is visualized using radiopaque markers under fluoroscopy, as shown in Fig. 5. The IVUS catheter has two radiopaque markers: one near the distal end and the other near the transducer. The former shows the location of the distal tip of the catheter in the vessel while the latter shows the location of IVUS observation. The size and the material of the markers need to be appropriately designed in view of not only visualization but also catheter configuration. In addition, the depth marker helps operator to recognize whether the tip of a catheter passed through a guiding catheter.

### 2.2.5. Safety

Safety is the most important aspect of medical devices. Not only biocompatibility but also fail-safe design is necessary to develop an IVUS catheter. In an IVUS-guided intervention, the proximal edge of the guidewire rail may rarely become stuck against the stent strut after stent placement. Although it is possible to release a catheter stuck in this manner by manipulating the catheter appropriately, it may be difficult due to the lack of stiffness in the imaging window section. A bailout measure has thus been implemented: the imaging core can be replaced with a stiffer guidewire to improve the catheter's flexural rigidity, as shown in Fig. 6.[16]

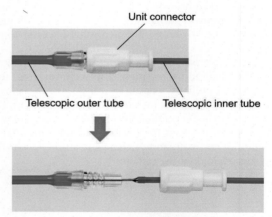

FIG. 6 Removal of imaging core for replacement with stiffer guidewire to enable release of stuck catheter.

### 2.3. Requirements for Ultrasound Probe

An IVUS diagnostic system obtains vascular images by scanning radially and longitudinally, as shown in Fig. 7. The characteristics of an IVUS differ from those of a conventional diagnostic ultrasound in terms of both the scanning method and the probe configuration, as shown in Fig. 8. In the following sections, we describe the structural features of the IVUS catheter used for acquiring ultrasound echo images of vascular lesions with regard to the ultrasound emission method and the transmission path.

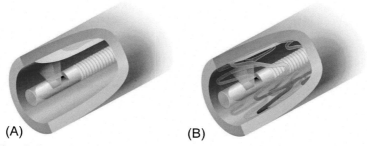

FIG. 7 Vessel observation from within by IVUS catheter: (A) before and (B) after stent placement.

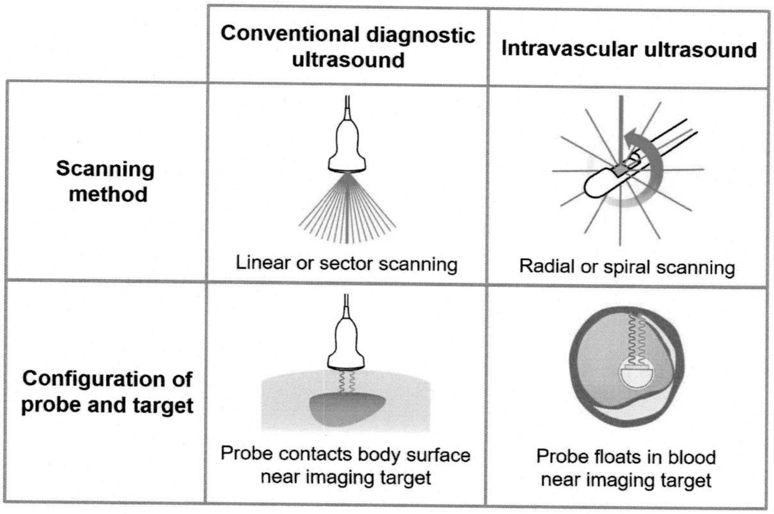

|  | Conventional diagnostic ultrasound | Intravascular ultrasound |
|---|---|---|
| **Scanning method** | Linear or sector scanning | Radial or spiral scanning |
| **Configuration of probe and target** | Probe contacts body surface near imaging target | Probe floats in blood near imaging target |

FIG. 8 Characteristics of IVUS system compared to those of conventional diagnostic ultrasound probe.

### 2.3.1. Radial imaging

There are two basic types of ultrasound transmission and reception methods for IVUS catheters: mechanical scanning and electronic phased array scanning, as shown in Fig. 9. With mechanical scanning, a single transducer rotates inside a catheter sheath via a flexible coil shaft spun by the motor of the MDU. With electronic phased array scanning, individual elements in a circumferential array of transducer elements mounted near the tip of the catheter are activated with different time delays to scan radially.[17]

Since the electrical signal converted from the ultrasound by the transducer is weak, the signal line between the transducer and the preamp needs to be shielded from ambient electromagnetic noises. With mechanical transmission, a flexible coil shaft is used for torque transmission as well as for the driving/pullback force

from the proximal to distal end in the catheter. An imaging core rotates within a catheter sheath to prevent direct contact between the imaging core and the blood vessel, and to stabilize the rotation of the flexible imaging core. Since a catheter bends along with the blood vessel running, the inner wall of the catheter sheath and the outer wall of the imaging core make contact following such bending. If the friction generated between the catheter sheath and the imaging core reaches a certain level, rotation transmission is hindered, which may cause nonuniform rotational distortion (NURD).

### 2.3.2. Longitudinal imaging

Longitudinal information about the blood vessel can be acquired by moving the entire catheter along the blood vessel running. In addition, with mechanical IVUS, information can be acquired in the longitudinal direction by moving only the imaging core relative to the sheath by utilizing the telescopic inner/outer tubes, which enables the sheath to remain in the blood vessel during pullback.

Although vascular conditions in terms of shape and elasticity differ between patients and locations, the sheath enables reproducible IVUS observation even under inhomogeneous conditions. The pullback distance is typically 150 mm, corresponding to the length of the coronary artery.

### 2.3.3. Ultrasound transmission

Since ultrasound is reflected or scattered at the boundary between materials with different acoustic impedances, the acoustic properties of the imaging window of the IVUS catheter should be taken into consideration. In addition, the void in an IVUS catheter needs to be filled

with heparinized saline as acoustic medium using the flushing procedure described in Section 1.2. This heparinized saline passes through the IVUS catheter from the proximal end to the distal end, as shown in Fig. 10.

## 3. SYSTEM CONFIGURATION
### 3.1. Motor Drive Unit

As mentioned in the introduction, the MDU plays important roles not only in rotating the transducer for radial scanning but also for moving it from the distal to proximal position at a constant speed for longitudinal scanning. Specifically, the scanner of the MDU moves on the base, as shown in Fig. 11. Precise control of two actuators for each scan underpins sophisticated

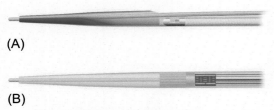

(A)

(B)

FIG. 9 (A) Mechanical IVUS and (B) electronic phased array IVUS.

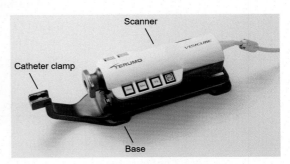

FIG. 11 Appearance of MDU (VISICUBE).

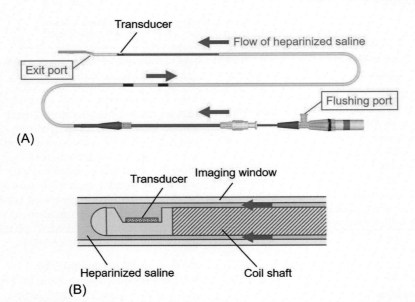

FIG. 10 (A) Route for heparinized saline within IVUS catheter. Heparinized saline flows within IVUS catheter from flushing port to exit port. (B) Enlarged cross section of IVUS imaging window section during flushing procedure.

IVUS image construction. For radial scanning, the MDU rotates the transducer via a coil shaft using a motor. Rotation speeds have now reached 90 rpm, which enables high-speed pullback. It is noteworthy that the MDU has functions for both electrical signal input/output and the movement of IVUS catheter at the same time.

Table 3 lists the specifications of a state-of-the-art IVUS diagnostic system (AltaView and VISICUBE). For longitudinal scanning, auto pullback scanning (up to 150 mm) is possible due to high-speed pullback. The pullback time of a conventional system is typically 300 s due to a low pullback speed of up to ~0.5 mm/s. Recent improvement in the pullback speed (up to 9.0 mm/s) has shortened the pullback time to 17 s. Since catheter intervention can cause ischemia, especially in the case of a diffuse lesion, a shorter pullback time is highly desirable. In addition, a shorter pullback time helps to reduce motion artifacts caused by heartbeat.

As described in Section 1.1, an MDU is a user interface of IVUS in the clean field. As shown in Fig. 11, the MDU has several switches, which enables operators to control pullback/hold the scanner and data acquisition in the clean field. In addition, an operator can manually slide the scanner. The length of a lesion is thus measureable by manipulating the MDU after setting a reference position of a digital ruler at one end of the lesion. Since the MDU is often held in the operator's hand during catheter manipulation and pullback, it needs to be light (typically <1 kg) and easy to hold.

## 3.2. Console

The console processes the electrical signal converted from the ultrasound by the transducer and constructs ultrasound cross-sectional images. IVUS images are displayed not only on a console monitor but also on cath-lab monitors via a video output. Since an IVUS system enables real-time observation of a blood vessel, an operator can plan a treatment strategy in view of stenosis severity and/or tissue characteristics. Moreover, auto pullback scanning using the MDU enables a certain range of cross-sectional images of the blood vessel to be recorded. The recorded images are useful for quantitatively evaluating the target lesion, such as the area of plaque or lumen in the blood vessel. In addition, auto pullback scanning enables a longitudinal IVUS image to be constructed by stacking cross-sectional images. A longitudinal IVUS image is useful for measuring the length of the target lesion. The recorded image data can be saved on storage media or data server.

## 4. SIGNAL PROCESSING
### 4.1. Image Construction

The signal processing for IVUS image construction is diagrammed in Fig. 12. With mechanical IVUS, a single transducer transmits and receives ultrasound alternatively. The console produces a trigger signal with a period equal to transducer rotation followed by signal transmission in accordance with the trigger signal. The transmitted electrical signal is converted into ultrasound by the transducer, which also converts the reflected ultrasound signals into an electrical signal. This ultrasound transmission and reception is done repetitively during each rotation. Fig. 13 shows radial scanning and image construction in the IVUS system. The number of radial scanning lines is typically 512–2048.

The obtained signal is amplified, analog-to-digital converted, filtered, and then converted into a 256-level grayscale image in the case of B-mode (brightness mode), in which the brightness of each dot is determined by the amplitude of the ultrasound echo signal.

**TABLE 3**
Typical Specifications of IVUS Diagnostic System (AltaView and VISICUBE).

| Frame rate | Pullback speed (mm/s) | Slice thickness (mm) | 150-MM PULLBACK | |
|---|---|---|---|---|
| | | | Time (s) | Number of frames |
| 30 fps (1800 rpm) | 0.5 | 0.017 | 300 | 9000 |
| | 1.0 | 0.033 | 150 | 4500 |
| | 2.0 | 0.067 | 75 | 2250 |
| 60 fps (3600 rpm) | 3.0 | 0.050 | 50 | 3000 |
| | 6.0 | 0.100 | 25 | 1500 |
| 90 fps (5400 rpm) | 3.0 | 0.033 | 50 | 4500 |
| | 6.0 | 0.067 | 25 | 2250 |
| | 9.0 | 0.100 | 17 | 1500 |

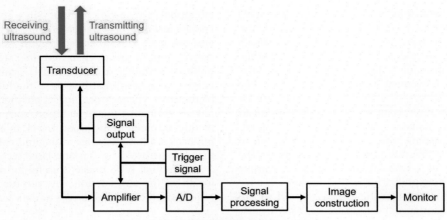

FIG. 12 Flowchart of IVUS signal processing.

B-mode is the most common output mode in ultrasonography because it provides easy-to-understand two-dimensional cross-sectional images in real time.

## 4.2. High-Frequency IVUS

The conventional frequency of IVUS signal has been about 10–40 MHz, but higher frequency IVUS (up to 60 MHz) has become common in recent years. Such high-frequency IVUS provides better image quality, as shown in Fig. 14. It has enabled better delineation with higher resolution and better contrast in tissue characteristics. Both axial and lateral spatial resolutions have been improved. The axial resolution is determined by the minimum distance between two target objects placed in the axial direction along which the two reflected ultrasound waves are distinguishable. Therefore, the axial resolution decreases as the ultrasound pulse width of the transmission wave becomes shorter, as shown in Fig. 15. Similarly, the lateral resolution is determined by the minimum distance between two target objects placed in the lateral direction along which the two reflected ultrasound waves are distinguishable. Since ultrasound beam width depends on ultrasound frequency, the lateral resolution improves as the ultrasound frequency increases, as shown in Fig. 16.

Ultrasound waves gradually attenuate due to the effects of absorption, scattering, and reflection as they propagate through a medium. Higher frequency ultrasound has lower penetration depth because the attenuation of the ultrasound amplitude depends not only on the depth but also on the frequency. In the case of VIS-ICUBE, an output frequency of 40–60 MHz can be selected to optimize resolution and penetration depth in accordance with the diameter of the blood vessel or purpose of the observation.

## 5. CONCLUDING REMARKS

In this chapter, we described an IVUS diagnostic system by focusing on the IVUS observation procedure, imaging catheter design concept, and signal processing for image construction. Improvements to the IVUS system since it was introduced almost 50 years ago have made it an important tool for image-guided intervention such as stent deployment. In addition, the need for IVUS in operator's decision-making has grown in response to a variety of intervention strategies. IVUS is thus currently an indispensable tool in catheter intervention especially in Japan.

On the other hand, simplification of the procedure would be as important as widely accepted requirements for IVUS such as image quality or pullback speed because the procedure of IVUS is very complex from preparation to image analysis. In response to this, we proposed the design concept for "Quick IVUS" with the emphasis on streamlining IVUS procedure such as preparation, delivery, observation, diagnosis, data management, and troubleshooting, as shown in Fig. 17.

Further improvements in both hardware and software will make IVUS an even more sophisticated tool. Development of a hybrid imaging catheter by fusing IVUS with another imaging modality will make IVUS suitable for clinical practice. Moreover, the clinical value of IVUS could be enhanced by combining it with cutting-edge digital technology such as artificial

## Single-line scanning

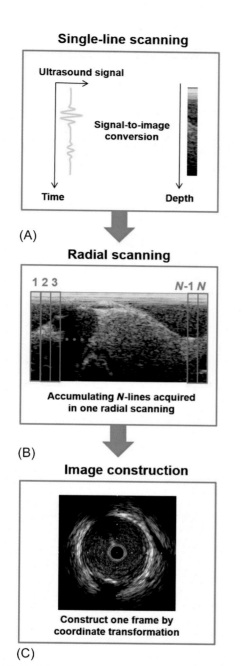

**(A)**

## Radial scanning

**Accumulating *N*-lines acquired
in one radial scanning**

**(B)**

## Image construction

**Construct one frame by
coordinate transformation**

**(C)**

FIG. 13 Radial scanning and image construction in IVUS system. An IVUS image is constructed by radial scanning of ultrasound transmission and reception. (A) Single-line scanning forms one radial imaging, where received ultrasound waveform corresponds to one-dimensional image. (B) Radial scanning enables to accumulate N-lines acquired in radial scanning. In this image, *vertical line* corresponds to radial direction (i.e., depth) while *horizontal line* corresponds to angle. (C) IVUS image is then constructed by coordinate transformation.

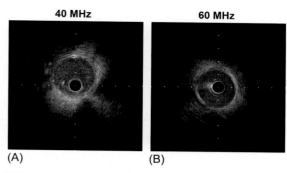

FIG. 14 Typical IVUS images with (A) 40-MHz and (B) 60-MHz signals.

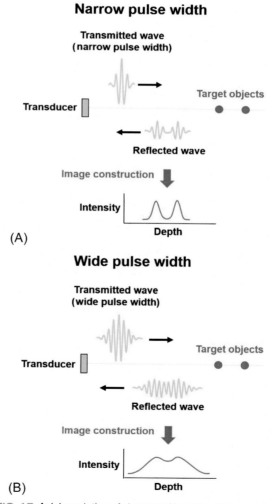

FIG. 15 Axial resolution of ultrasound imaging. Narrow pulse width of ultrasound wave (A) is better than wide pulse width (B) in terms of axial resolution.

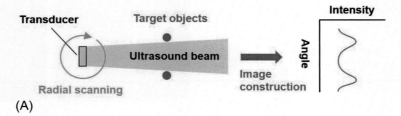

FIG. 16 Lateral resolution of ultrasound imaging. High frequency of ultrasound wave (A) is better than low frequency (B) in terms of lateral resolution.

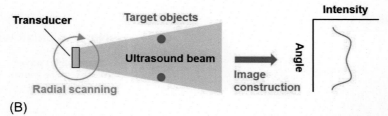

FIG. 17 "Quick IVUS" concept for streamlining IVUS procedure.

intelligence. This would enable it to assist in the decision-making process in the field of catheter intervention.

## ACKNOWLEDGMENT

The authors thank Hiroyuki Yagami, Ryota Sugimoto, Noboru Saito, Hiroshi Sato, Keiichiro Yamamoto, Tomoji Maruyama, Yasunori Yamashita, Yuji Hojo, Isao Mori, Soichiro Nakagiri, Kanae Hashimoto, and Misa Koike of Terumo Corporation, for their support and helpful discussions.

## REFERENCES

1. Bom N, ten Hoff H, Lancée CT, Gussenhoven WJ, Bosch JG. Early and recent intraluminal ultrasound devices. *Int J Card Imaging*. 1989; 4:79–88.
2. Yock PG, Linker DT, Angelsen BA. Two-dimensional intravascular ultrasound: technical development and initial clinical experience. *J Am Soc Echocardiogr*. 1989; 2:296–304.
3. Yock PG, Fitzgerald PJ, Linker DT, Angelsen BA. Intravascular ultrasound guidance for catheter-based coronary interventions. *J Am Coll Cardiol*. 1991; 17:39B–45B.
4. Honye J, Mahon DJ, Jain A, et al. Morphological effects of coronary balloon angioplasty in vivo assessed by intravascular ultrasound imaging. *Circulation*. 1992; 85 (3):1012–1025.
5. Nissen SE, Yock P. Intravascular ultrasound: novel pathophysiological insights and current clinical applications. *Circulation*. 2001; 103(4):604–616.
6. Mintz GS, Nissen SE, Anderson WD, et al. American College of Cardiology Clinical Expert Consensus Document on Standards for Acquisition, Measurement and Reporting of Intravascular Ultrasound Studies

(IVUS). A report of the American College of Cardiology Task Force on Clinical Expert Consensus Documents. *J Am Coll Cardiol.* 2001; 37(5):1478–1492.

7. Mintz GS. Intravascular imaging of coronary calcification and its clinical implications. *JACC Cardiovasc Imaging.* 2015; 8(4):461–471.

8. Mintz GS, Guagliumi G. Intravascular imaging in coronary artery disease. *Lancet.* 2017; 390(10096):793–809.

9. Räber L, Mintz GS, Koskinas KC, et al. Clinical use of intracoronary imaging. Part 1: guidance and optimization of coronary interventions. An expert consensus document of the European Association of Percutaneous Cardiovascular Interventions. *Eur Heart J.* 2018; 39(35):3281–3300.

10. Sonoda S, Hibi K, Okura H, Fujii K, Honda Y, Kobayashi Y. Current clinical use of intravascular ultrasound imaging to guide percutaneous coronary interventions. *Cardiovasc Interv Ther.* 2020; 35(1):30–36.

11. Fitzgerald PJ, Oshima A, Hayase M, et al. Final results of the Can Routine Ultrasound Influence Stent Expansion (CRUISE) study. *Circulation.* 2000; 102(5):523–530.

12. Levine GN, Bates ER, Blankenship JC, et al. American College of Cardiology Foundation; American Heart Association Task Force on Practice Guidelines; Society for Cardiovascular Angiography and Interventions. 2011 ACCF/AHA/SCAI Guideline for Percutaneous Coronary Intervention. A report of the American College of Cardiology Foundation/American Heart Association Task Force on Practice Guidelines and the Society for Cardiovascular Angiography and Interventions. *J Am Coll Cardiol.* 2011; 58:e44–122.

13. Neumann FJ, Sousa-Uva M, Ahlsson A, et al. 2018 ESC/EACTS guidelines on myocardial revascularization. *Eur Heart J.* 2019; 40:87–165.

14. Villa AD, Sammut E, Nair A, Rajani R, Bonamini R, Chiribiri A. Coronary artery anomalies overview: the normal and the abnormal. *World J Radiol.* 2016; 8(6):537–555.

15. Okamura A, Iwakura K, Date M, Nagai H, Sumiyoshi A, Fujii K. Navifocus WR is the promising intravascular ultrasound for navigating the guidewire into true lumen during the coronary intervention for chronic total occlusion. *Cardiovasc Interv Ther.* 2014; 29(2):181–186.

16. Okubo I, Mitsuhashi K, Sakaguchi Y. US Patent 9931101; 2018.

17. Kitahara H, Honda Y, Fitzgerald PJ. Intravascular ultrasound. In: Lanzer P, ed. *PanVascular Medicine.* Berlin, Heidelberg: Springer; 2015:1379–1418.

# CHAPTER 7

# Multimodality Intravascular Imaging Technology

ERIC A. OSBORN[a] • GIOVANNI J. UGHI[b]
[a]Cardiovascular Division, Beth Israel Deaconess Medical Center, Harvard Medical School, Boston, MA, United States, [b]New England Center for Stroke Research, Department of Radiology, University of Massachusetts Medical School, Worcester, MA, United States

## 1. INTRODUCTION

IVUS imaging was introduced clinically in the early 1990s and enabled the detection of plaque and lumen characteristics in living subjects at levels not before achievable with traditional contrast angiography. The first clinical application of intravascular imaging techniques sparked significant interest in investigating detection schemes to identify a subset of atherosclerotic plaques that are at high risk to cause clinical events.[1] Catheter-based intravascular imaging is routinely adopted today in clinical practice, where it is used as an adjunct to coronary angiography to precisely plan and then optimize the result of percutaneous coronary interventions (PCI). IVUS and optical coherence tomography (OCT) are the primary intravascular imaging tools available for use in the cardiac catheterization laboratory. They are often referred to as "structural" imaging techniques, as they provide detailed information about the vessel lumen and wall morphology but offer little insight into the underlying plaque pathobiology. Both IVUS and OCT are utilized similarly during PCI procedures: (1) the pre-PCI planning stage sizes the stent appropriately to the vessel, and (2) the post-PCI optimization stage ensures complete stent expansion and apposition to the vessel wall and evaluates for any stent-related vascular compromise. Inadequate stent expansion, arterial wall dissections, and thrombus represent key imaging parameters correlated with adverse events that may not be appreciated by angiography, but can be reliably detected by intravascular imaging.[2] Recent studies and metaanalyses compared intravascular imaging-guided PCI to angiography alone, demonstrating a clear benefit for image-guided interventions in terms of the need for repeat revascularization and major adverse events.[3–5]

Despite their established role as tools to optimize coronary treatments, recent large-scale clinical trials and histology-based studies have exposed multiple limitations of structural intravascular imaging platforms in the detection of high-risk ("vulnerable") plaque characteristics.[6–9] These investigations have thus uncovered an unmet need to develop alternative invasive imaging modalities that can more accurately detect plaque chemical composition and cellular and molecular pathobiology in vivo. Recent technological advances, including probe miniaturization as well as more efficient tools for data processing, have enabled the development of novel imaging modalities including spectroscopy, fluorescence, and photoacoustics methods. Given their ability to provide complementary information about plaque composition and its functional state, these new and emerging intravascular imaging may overcome the limitations of stand-alone IVUS and OCT structural imaging. Fundamentally, no single image modality can provide a complete assessment of the complex disease state in atherosclerosis, and the development of multimodality approaches has been an intense topic of research and clinical development in the last decade. In this chapter, we review the recent developments for multimodality coronary artery imaging, summarizing their current status, including technology progression and validation.

## 2. HYBRID IVUS-OCT IMAGING TECHNOLOGY

Optical coherence tomography (OCT) is a clinical imaging modality that uses near-infrared light to generate high-resolution tomographic images of the wall and the inner lumen of coronary arteries.[10,11] Similar to IVUS, an OCT catheter includes a short, rapid exchange tip designed to navigate the catheter through blood vessels over a 0.014" coronary guidewire using standard clinical techniques. As blood is a highly scattering media in the near-infrared wavelengths, the acquisition of OCT

*Intravascular Ultrasound.* https://doi.org/10.1016/B978-0-12-818833-0.00007-2

data requires flushing to temporarily displace blood from the arterial lumen. Flushing is typically performed by means of a 10–15 mL iodinated contrast injection through an appropriately engaged guiding catheter, the amount used for a typical coronary angiogram cine acquisition. Although contrast provides the most consistent blood displacement due to its higher viscosity, in some instances a 0.9% saline solution (with or without contrast dilution) is utilized as a contrast-sparing flushing technique. Over a duration of approximately 3 s during flushing, the OCT catheter rotating optical lens is automatically retracted at pullback speeds of 20 mm/s or faster, rapidly generating a volumetric data set from a coronary segment longer than 50 mm. Currently, state-of-the-art commercial systems acquire OCT data at a frame rate of approximately 180 cross-sectional images/s. The use of OCT imaging has been extensively investigated over the last decade showing a very high safety profile equivalent to IVUS,[12] and improved procedural outcomes compared to standard angiography-guided PCI.[3]

In comparison to IVUS-generated sound wave images, the shorter wavelength of light-based OCT imaging achieves significant improvements in spatial resolution: 15 μm versus approximately 100 μm. As such, the higher resolution of OCT offers several advantages for the visualization of intraluminal objects such as vessel dissections, coronary stents, and thrombosis (Fig. 1),[13–15] as well as for the quantification of lumen dimensions and stent sizing.[16] IVUS offers greater image penetration depth over OCT in lipid-rich and necrotic plaques, enabling characterization and quantification of the atherosclerotic plaque burden important for assessing plaque growth or regression in pharmacological studies (Fig. 2). However, IVUS ultrasound waves exhibit shallow penetration through calcific tissue due to acoustic reflections, and therefore are unable to quantify and characterize the thickness of coronary calcifications. On the contrary, near-infrared light is readily capable of transmitting through calcium with low attenuation,[17,18] and OCT images can be used to quantify and depict in high-resolution the coronary calcification microstructure, angular distribution, and thickness. OCT calcium evaluation thus provides a comprehensive assessment of plaque calcification which can be used clinically to determine the need for coronary atherectomy or other plaque modification techniques prior to stent implantation (Fig. 3).

Given the complementary nature of these imaging modalities, a significant body of research has focused on developing methods for merging IVUS and OCT into a hybrid imaging catheter.[19,20] Feasibility for in vivo combination IVUS-OCT imaging has been demonstrated,[21] and different solutions have been proposed to acquire co-registered images and eliminate rotational or longitudinal offsets, a frequent challenge in the development of multimodality approaches. Recently, a device with a profile of approximately 3F (1 mm diameter) obtained regulatory clearance and completed the translation of a hybrid IVUS-OCT probe to the clinic. In clinical imaging scenarios, hybrid IVUS-OCT imaging applications illustrate some of the complimentary aspects of these modalities for the characterization of coronary artery disease and stent implantation.[22] As anticipated, based on their stand-alone counterparts, vessel dissections, calcific lesions, and stent apposition were better appreciated via OCT, while deeper features such as plaque burden were more accurately assessed by IVUS.

## 3. NEAR-INFRARED SPECTROSCOPY

In an effort to increase the specificity of structural imaging modalities to identify lipid-rich plaques, near-infrared spectroscopy (NIRS) imaging has been combined with IVUS in a device approved for clinical use (Fig. 4).[23,24] Prototypes of OCT-NIRS imaging systems are also currently under development.[25] The fundamental principle of NIRS is to use diffuse near-infrared light to interrogate the chemical characteristics of plaque constituents within the coronary wall. By analyzing the light absorption spectrum of cholesterol moieties, specific plaque lipid chemical signatures are identified allowing for an enhanced detection of lipid core plaques (LCP). NIRS catheter technology was validated in histopathology studies and shown to accurately identify lipid core features in ex vivo coronary plaques.[26] Multiple studies have illustrated the feasibility of a commercially available IVUS-NIRS catheter to reliably collect spectroscopy data in human patients for the invasive detection of LCP.[27] Early NIRS clinical studies demonstrated that unstable culprit lesions exhibit greater LCP,[28] and that detection of LCP zones prior to coronary intervention was predictive of the risk of peri-procedural myocardial infarction.[29] In a recent prospective clinical trial, IVUS-NIRS imaging of non-culprit coronary lesions was performed in 1563 patients undergoing PCI to establish whether IVUS-NIRS could predict locations at risk for future adverse cardiac events.[30] In this landmark study, the use of an NIRS-derived metric termed the Lipid Core Burden Index (LCBI) was beneficial in identifying patients and coronary lesion locations with a higher risk for

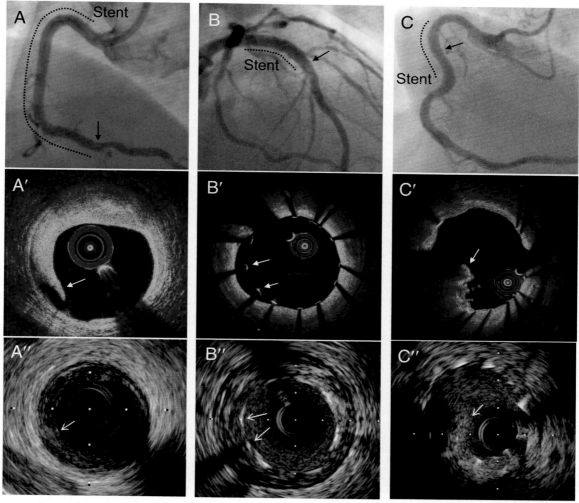

FIG. 1 Comparison of OCT and IVUS for the assessment of intracoronary stents in human patients. (A) OCT and IVUS images of a coronary artery dissection flap *(white arrows)*, showing the benefits of the higher resolution provided by the use of light-based OCT imaging, (B) incomplete stent apposition to the coronary wall *(white arrows)*, and (C) an example of intraluminal thrombus *(white arrows)*. Notably, none of the findings identified by OCT *(middle row)* or IVUS *(bottom row)* intracoronary imaging is apparent on the corresponding X-ray coronary angiography images *(top row)*. (Reproduced with permission from Maehara A, Matsumura M, Ali ZA, Mintz GS, Stone GW. IVUS-guided versus OCT-guided coronary stent implantation: a critical appraisal. *J Am Coll Cardiol Img.* 2017;10(12):1487-1503.)

subsequent coronary events at sites that were remote from the primary lesion. These data support the ability of intracoronary NIRS chemical lipid sensing to identify future atherosclerosis risk in patients and to motivate additional studies to investigate antiatherosclerosis pharmaceutical targets that may offer strategies to reduce this risk.

## 4. INTRAVASCULAR PHOTOACOUSTICS

Several research groups have investigated the use of intravascular photoacoustics (IVPA) as an alternative tool to characterize coronary plaque structure and composition. As a natural extension to an IVUS imaging catheter, sound-derived photoacoustics measures hold potential to add tissue-type specificity to US-generated

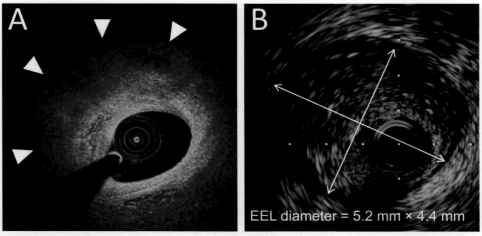

FIG. 2 OCT (A) and IVUS (B) imaging of a lipid-rich coronary artery plaque. Compared to near-infrared light, the deeper penetration of ultrasound waves through lipid plaques allows a precise assessment of the plaque burden and the external elastic lamina (EEL) (B). The higher resolution OCT image enables a more detailed characterization of the microstructure of the fibrotic cap covering the deeper plaque features (A). (Reproduced with permission from Maehara A, Matsumura M, Ali ZA, Mintz GS, Stone GW. IVUS-guided versus OCT-guided coronary stent implantation: a critical appraisal. *J Am Coll Cardiol Img.* 2017;10(12):1487-1503.)

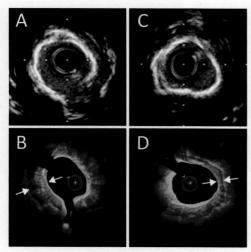

FIG. 3 IVUS (A, C) and OCT (B, D) assessment of coronary calcifications. Near-infrared light utilized to generate OCT images can penetrate deeper into calcific tissue than ultrasound waves, which are highly reflected by dense calcium. OCT, therefore, allows a more complete characterization of plaque calcium deposits than IVUS, by enabling quantification of the thickness of the coronary calcification. Such information can be used to guide coronary intervention techniques, such as the need for atherectomy procedures in order to optimize lesion preparation prior to stent implantation. (Reproduced with permission from Mintz GS, Guagliumi G. Intravascular imaging in coronary artery disease. *Lancet.* 2017;390 (10096):793-809.)

intracoronary images. By acquiring the ultrasonic signal emitted from the tissue exposed to short, high-energy light pulses, IVPA analyzes the differences in the absorption spectra of the various tissue components present in order to reconstruct the tissue composition of the coronary wall. The spatial resolution of acquired IVPA imaging data is comparable to that of IVUS, and the IVPA signal can be further depth resolved to visualize key plaque features such as the location of lipid accumulations within the coronary wall (Fig. 5).[31] In addition, by tuning the optimal excitation light wavelength, IVPA can probe specific tissue components. For example, two IVPA absorption bands localized at 1210 and 1720 nm result from C—H bond vibrations within lipid molecules.[32] By selecting wavelengths centered around these values, recent studies have reported a successful discrimination of different types of lipid accumulations within the coronary wall, facilitating differentiation between pathological plaque lipid deposits and benign perivascular fat.[33,34]

Dual-modality IVUS-IVPA probes are designed to combine in a single device an ultrasound transducer and the optical fibers required for laser tissue excitation. In preclinical feasibility studies, flexible catheters with a profile size compatible for coronary imaging have been demonstrated in vivo.[35-37] However, multiple challenges in the development of IVPA intracoronary devices remain, including: (1) limited imaging acquisition speed (20 frames/s), (2) requirement for flushing with contrast or saline to clear the arterial lumen from blood,

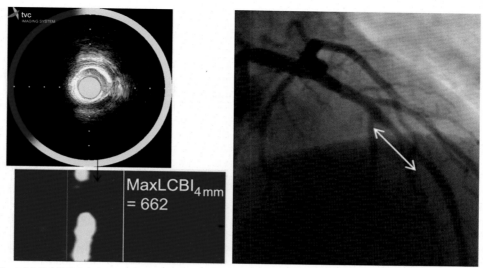

**FIG. 4** IVUS-NIRS imaging of the LAD coronary artery in a patient undergoing cardiac catheterization. The lesion was characterized by IVUS and NIRS as a calcified lipid-rich plaque. Given the limited penetration of ultrasound through calcium, a stand-alone IVUS image would have been restricted to assessing the superficial plaque. The two-dimensional NIRS chemogram (*bottom left image*; *x*-axis = distance, *y*-axis = rotational angle) visualizes the distribution of lipid deposits (*yellow*) in the arterial segment under investigation on the coronary angiogram *(right image; double-head white arrow)*. Intracoronary NIRS data are also displayed as a color-coded ring band placed around the cross-sectional IVUS image *(top left image)*: *red* color indicates a low probability of lipid accumulations and *yellow* a higher probability. (Reproduced with permission from Stone GW, Maehara A, Muller JE, et al. Plaque characterization to inform the prediction and prevention of periprocedural myocardial infarction during percutaneous coronary intervention: the CANARY trial (coronary assessment by near-infrared of atherosclerotic rupture-prone yellow). *JACC Cardiovasc Interv.* 2015;8(7):927-936.)

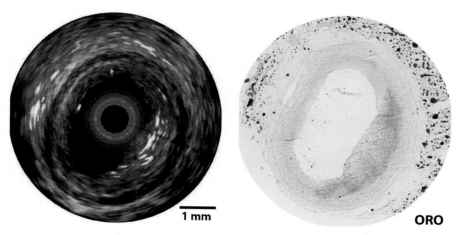

**FIG. 5** Ex vivo imaging of human atherosclerotic plaque using an IVUS-IVPA dual-imaging probe *(left image)*. The IVUS (grayscale intensity) image is combined with IVPA data by overlaying the two images. *Yellow* indicates the presence of lipid plaque components (3–4 o'clock). The *image on the right* shows the corresponding histological cross section stained using ORO for neutral lipids *(red color)*, corroborating the IVPA findings of a lipid-rich plaque in the *bottom right* quadrant. (Reproduced with permission from Wu M, Springeling G, Lovrak M, et al. Real-time volumetric lipid imaging in vivo by intravascular photoacoustics at 20 frames per second. *Biomed Opt Express.* 2017;8(2):943-953 © The Optical Society.)

(3) restrictions on the optical laser power resulting in weaker IVPA-induced ultrasound signals than IVUS, and (4) available catheter materials transparent to both light and ultrasound waves. This complex set of contravening engineering specifications has slowed the translation of IVPA imaging modalities to the clinic.

## 5. NEAR-INFRARED FLUORESCENCE

Structural imaging techniques such as IVUS and OCT are essential tools to depict the lumen morphology and vessel wall microstructure in great detail. However, it has become increasing apparent that structural imaging alone lacks capacity to capture the complex biological underpinnings that drive atherosclerotic plaque progression, as well as decipher the risk for complications related to local plaque instability or coronary stent failure. To overcome the inherent limitations of structural imaging, near-infrared fluorescence (NIRF) intravascular molecular imaging has been proposed as a highly specific method to visualize and characterize in vivo plaque and stent biology, including important areas such as coronary artery inflammation, atherosclerosis disease progression, high-risk plaque features, and device prothrombotic milieu.[38] The fundamental principle of NIRF molecular imaging is to label cellular and molecular targets of interest with NIR fluorochromes. Fluorochromes are administered through an IV injection to subjects, they co-localize within tissues and report on a biological process of interest in vivo. For detection, catheter-based NIRF imaging systems excite labeled tissue using a specific light wavelength in the NIR that both minimizes tissue autofluorescence and allows improved tissue depth penetration over visible light wavelengths, and then the same imaging catheter collects the red-shifted fluorescence signal emitted from the excited fluorochrome. As NIRF is solely a biological imaging platform without structural information, NIRF has been combined with OCT or IVUS to create clinically translatable dual-modality imaging systems.[39–41] Hybrid OCT-NIRF devices are similar to conventional OCT catheters, with the exception of replacing the use of a single-mode fiber with custom made dual-clad fibers. In the dual-clad fiber configuration, a single-mode fiber core is used for the acquisition of structural OCT images, and the fiber clad with a larger optical numerical aperture (NA) is used for the collection of the emitted fluorescence signal for molecular NIRF imaging.

In a preclinical in vivo model, OCT-NIRF dual-modality catheters were used in combination with intravenous injections of either inflammation or fibrin-targeted NIRF molecular imaging agents, showing their potential to capture molecular information in the context of tissue and lumen microstructure.[39] Different types of NIRF molecular imaging agents have been explored for intravascular imaging, with a focus on clinically translatable probes. One promising candidate is indocyanine green (ICG), an FDA-approved NIR fluorescent dye that exhibits excitation and emission wavelengths of 740–800 nm and 800–860 nm, respectively. ICG has been characterized for the detection of atherosclerosis and validated in a series of histopathology studies in preclinical settings, revealing co-localization with resident macrophages and lipid-rich regions within atherosclerotic plaques.[38] In human carotid atherosclerosis subjects, ICG injected intravenously prior to planned surgical plaque resection enabled targeted NIRF molecular imaging of plaque macrophages, lipids, and intraplaque hemorrhage, and therefore may enable reporting on a subset of atheroma with high-risk features in patients.[42] Other experimental NIRF agents have been explored preclinically, with feasibility demonstrated for the characterization of atherosclerotic plaque inflammation using a cysteine protease NIRF reporter (ProSense VM110; PerkinElmer) and detection of fibrin-rich thrombus with a previously validated clinical FTP11 fibrin-targeted peptide,[43] labeled with a CyAm7 NIR fluorochrome.[39,44] Preclinical investigations have also shown that OCT-NIRF can identify areas of fibrin deposition within coronary stents that are associated with the absence of a protective endothelial layer, thus representing pathobiological markers of abnormal stent healing that may better identify stents at a heightened risk of thrombosis.[45] Tissue NIR autofluorescence (NIRAF), excited at 633 nm, has been proposed as a label-free alternative method to obtain in vivo biological information. In histopathology studies, NIRAF can function as an endogenous imaging biomarker and capture high-risk features of atherosclerotic plaque progression such as intra-plaque hemorrhage and oxidative stress.[46–48] The safety and efficacy of a dual modality OCT-NIRAF catheter has been established in human patients in a first-in-man study completing the translation of future clinical NIRF molecular imaging systems.[41] These initial results illustrate how the use of an endogenous imaging biomarker identified by NIRAF can provide complimentary biological information to the structural plaque data obtained from stand-alone OCT and IVUS images (Fig. 6).

## 6. FLUORESCENCE LIFETIME IMAGING

Fluorescence lifetime imaging (FLIM) has been introduced as an alternative to NIRAF for label-free

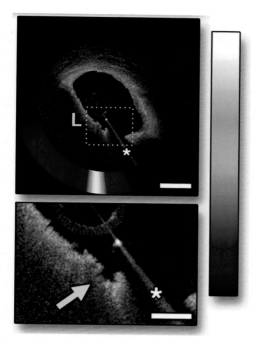

FIG. 6 OCT-NIRAF imaging of a plaque rupture located in the ostial LAD of a patient undergoing PCI *(upper image)*. Autofluorescence intensity data are displayed as a ring band surrounding the OCT cross-sectional image. *Blue* indicates low fluorescence emission and *green* to *white* color a higher NIRAF signal intensity. The rupture site of a lipid-rich plaque *(arrow in lower image corresponding to dashed box in upper image)* is co-localized with an area of elevated and focal autofluorescence signal *(green/white color band in upper image)*. The asterisk indicates the optical shadow trailed by the guidewire. (Reproduced with permission from Ughi GJ, Wang H, Gerbaud E, et al. Clinical characterization of coronary atherosclerosis with dual-modality OCT and near-infrared autofluorescence imaging. *J Am Coll Cardiol Img*. 2016;9(11):1304-1314.)

molecular imaging capabilities, complementing IVUS or OCT technology with information about the plaque biochemical composition. By exciting the vessel wall with UV light, several different tissue constituents exhibit autofluorescence. An optimized characterization of fluorescence lifetime signals has been obtained using a multispectral approach, splitting the collected signal into separate fluorescence channels between 300 and 700 nm to detect different tissue components.[49] Ex vivo studies demonstrated that the lifetime of the fluorescence state for different wavelength bands enables the detection of elastin, collagen, and lipid components of plaques and macrophages.[50] A limitation of FLIM

imaging, however, is that the estimated penetration of UV light into the vessel wall is only approximately 200 μm, and therefore FLIM is primarily able to interrogate the superficial intimal layer and the fibrotic cap covering atherosclerotic plaques. Despite the limited penetration, FLIM imaging can provide complementary information to structural OCT or IVUS regarding the inflammatory state and mechanical stability of the fibrotic cap, by assessing the ratio between collagen and lipid components.[50] Following these initial histopathology investigations, an IVUS-FLIM rotational catheter has been developed and validated in a combination of ex vivo and in vivo preclinical models.[51] Combining FLIM with OCT technology in a hybrid intravascular imaging system has also been proposed as an alternative.[52] Similar to stand-alone intravascular devices, catheter-based FLIM imaging can perform a volumetric acquisition of an arterial segment by means of a pullback, holding the potential to detect the biochemical features of multiple coronary plaque tissue types in the context of their morphology within a single catheter pullback.

## 7. MULTIMODALITY IMAGING: CLINICAL OUTLOOK

Available stand-alone intracoronary imaging modalities each exhibit strengths and weaknesses specific to the technology employed. Recent advances in biology, biophotonics, signal processing, and miniaturization of medical devices have enabled the development of an increasing number of multimodality imaging catheters designed to overcome the limitations of stand-alone imaging systems and combine the benefits of each imaging type into a single device. Several studies have demonstrated that multimodality imaging systems are yielding new and important insights into high-risk atherosclerotic plaque features in vivo. Based on promising preclinical validation findings, multimodality intracoronary imaging may offer a more comprehensive assessment of coronary plaque pathological mechanisms associated with adverse coronary events. Recent prospective clinical trials illustrated the potential of NIRS-IVUS intracoronary imaging to predict future events based on identifying lipid-rich plaque characteristics in nonculprit lesions.[30] Emerging intracoronary multimodality NIRF imaging catheters will offer new approaches to interrogate plaque biology in vivo. Nevertheless, significant challenges remain related to the implementation of multimodality imaging systems for patient care. The inherently invasive nature of intravascular imaging, a higher cost associated with purchasing the imaging system and catheters, the added complexity of the knowledge base required to interpret the findings, and the inability to visualize the pathophysiology of the entire coronary tree are fundamental drawbacks that may limit the use of hybrid modalities in a broad population. In addition, most of the abovementioned modalities are at their infancy, and the limited evidence base at present to support clinical use will need to be significantly expanded. Despite these challenges, supplementing IVUS and OCT with spectroscopic and molecular imaging modalities presents a unique opportunity for the study of coronary artery disease. Overall, multimodality imaging has the potential to expand the boundaries of stand-alone morphological imaging and provide unique insights into the pathobiology of complex coronary lesions. Further effort is anticipated in the next decade to translate and adopt hybrid coronary imaging in clinical practice and research in an effort to improve the identification and treatment of high-risk patients and atherosclerotic plaques.

# REFERENCES

1. Stone GW, Maehara A, Lansky AJ, et al. A prospective natural-history study of coronary atherosclerosis. *N Engl J Med.* 2011; 364(3):226–235.
2. Koskinas KC, Ughi GJ, Windecker S, Tearney GJ, Raber L. Intracoronary imaging of coronary atherosclerosis: validation for diagnosis, prognosis and treatment. *Eur Heart J.* 2016; 37(6):524–535a-c.
3. Jones DA, Rathod KS, Koganti S, et al. Angiography alone versus angiography plus optical coherence tomography to guide percutaneous coronary intervention: outcomes from the Pan-London PCI Cohort. *JACC Cardiovasc Interv.* 2018; 11(14):1313–1321.
4. Hong SJ, Kim BK, Shin DH, et al. Effect of intravascular ultrasound-guided vs angiography-guided everolimus-eluting stent implantation: the IVUS-XPL randomized clinical trial. *JAMA.* 2015; 314(20):2155–2163.
5. Zhang J, Gao X, Kan J, et al. Intravascular ultrasound versus angiography-guided drug-eluting stent implantation: the ULTIMATE trial. *J Am Coll Cardiol.* 2018; 72(24): 3126–3137.
6. Kaul S, Diamond GA. Improved prospects for IVUS in identifying vulnerable plaques? *J Am Coll Cardiol Img.* 2012; 5(3 Suppl):S106–S110.
7. Ahmadi A, Stone GW, Leipsic J, et al. Prognostic determinants of coronary atherosclerosis in stable ischemic heart disease: anatomy, physiology, or morphology? *Circ Res.* 2016; 119(2):317–329.
8. Arbab-Zadeh A, Fuster V. The myth of the "vulnerable plaque": transitioning from a focus on individual lesions to atherosclerotic disease burden for coronary artery disease risk assessment. *J Am Coll Cardiol.* 2015; 65 (8):846–855.
9. Libby P, Pasterkamp G. Requiem for the 'vulnerable plaque'. *Eur Heart J.* 2015; 36(43):2984–2987.
10. Tearney GJ, Regar E, Akasaka T, et al. Consensus standards for acquisition, measurement, and reporting of intravascular optical coherence tomography studies: a report from the International Working Group for Intravascular Optical Coherence Tomography Standardization and Validation. *J Am Coll Cardiol.* 2012; 59(12):1058–1072.
11. Gounis MJ, Ughi GJ, Marosfoi M, et al. Intravascular optical coherence tomography for neurointerventional surgery. *Stroke.* 2018; https://doi.org/10.1161/STROKEAHA. 118.022315.
12. van der Sijde JN, Karanasos A, van Ditzhuijzen NS, et al. Safety of optical coherence tomography in daily practice: a comparison with intravascular ultrasound. *Eur Heart J Cardiovasc Imaging.* 2017; 18(4):467–474.
13. Gerbaud E, Weisz G, Tanaka A, et al. Multi-laboratory inter-institute reproducibility study of IVOCT and IVUS assessments using published consensus document definitions. *Eur Heart J Cardiovasc Imaging.* 2016; 17 (7):756–764.
14. Ughi GJ, Van Dyck CJ, Adriaenssens T, et al. Automatic assessment of stent neointimal coverage by intravascular optical coherence tomography. *Eur Heart J Cardiovasc Imaging.* 2014; 15(2):195–200.
15. De Cock D, Bennett J, Ughi GJ, et al. Healing course of acute vessel wall injury after drug-eluting stent implantation assessed by optical coherence tomography. *Eur Heart J Cardiovasc Imaging.* 2014; 15(7):800–809.
16. Ughi GJ, Adriaenssens T, Desmet W, D'Hooge J. Fully automatic three-dimensional visualization of intravascular optical coherence tomography images: methods and feasibility in vivo. *Biomed Opt Express.* 2012; 3(12): 3291–3303.
17. Xu C, Schmitt JM, Carlier SG, Virmani R. Characterization of atherosclerosis plaques by measuring both backscattering and attenuation coefficients in optical coherence tomography. *J Biomed Opt.* 2008; 13(3). 034003.
18. Ughi GJ, Adriaenssens T, Sinnaeve P, Desmet W, D'Hooge J. Automated tissue characterization of in vivo atherosclerotic plaques by intravascular optical coherence tomography images. *Biomed Opt Express.* 2013; 4(7): 1014–1030.
19. Li BH, Leung AS, Soong A, et al. Hybrid intravascular ultrasound and optical coherence tomography catheter for imaging of coronary atherosclerosis. *Catheter Cardiovasc Interv.* 2013; 81(3):494–507.
20. Yin J, Yang HC, Li X, et al. Integrated intravascular optical coherence tomography ultrasound imaging system. *J Biomed Opt.* 2010; 15(1). 010512.
21. Yin J, Li X, Jing J, et al. Novel combined miniature optical coherence tomography ultrasound probe for in vivo intravascular imaging. *J Biomed Opt.* 2011; 16(6). 060505.

22. Sheth TN, Pinilla-Echeverri N, Mehta SR, Courtney BK. First-in-human images of coronary atherosclerosis and coronary stents using a novel hybrid intravascular ultrasound and optical coherence tomographic catheter. *JACC Cardiovasc Interv.* 2018; 11(23):2427–2430.

23. Schuurman AS, Vroegindewey M, Kardys I, et al. Near-infrared spectroscopy-derived lipid core burden index predicts adverse cardiovascular outcome in patients with coronary artery disease during long-term follow-up. *Eur Heart J.* 2018; 39(4):295–302.

24. Madder RD, Husaini M, Davis AT, et al. Large lipid-rich coronary plaques detected by near-infrared spectroscopy at non-stented sites in the target artery identify patients likely to experience future major adverse cardiovascular events. *Eur Heart J Cardiovasc Imaging.* 2016; 17(4): 393–399.

25. Fard AM, Vacas-Jacques P, Hamidi E, et al. Optical coherence tomography—near infrared spectroscopy system and catheter for intravascular imaging. *Opt Express.* 2013; 21 (25):30849–30858.

26. Gardner CM, Tan H, Hull EL, et al. Detection of lipid core coronary plaques in autopsy specimens with a novel catheter-based near-infrared spectroscopy system. *J Am Coll Cardiol Img.* 2008; 1(5):638–648.

27. Waxman S, Dixon SR, L'Allier P, et al. In vivo validation of a catheter-based near-infrared spectroscopy system for detection of lipid core coronary plaques: initial results of the SPECTACL study. *J Am Coll Cardiol Img.* 2009; 2(7): 858–868.

28. Madder RD, Smith JL, Dixon SR, Goldstein JA. Composition of target lesions by near-infrared spectroscopy in patients with acute coronary syndrome versus stable angina. *Circ Cardiovasc Interv.* 2012; 5(1):55–61.

29. Goldstein JA, Maini B, Dixon SR, et al. Detection of lipid-core plaques by intracoronary near-infrared spectroscopy identifies high risk of periprocedural myocardial infarction. *Circ Cardiovasc Interv.* 2011; 4(5):429–437.

30. Waksman R, Di Mario C, Torguson R, et al. Identification of patients and plaques vulnerable to future coronary events with near-infrared spectroscopy intravascular ultrasound imaging: a prospective, cohort study. *Lancet.* 2019.

31. Wu M, Springeling G, Lovrak M, et al. Real-time volumetric lipid imaging in vivo by intravascular photoacoustics at 20 frames per second. *Biomed Opt Express.* 2017; 8(2): 943–953.

32. Jansen K, Wu M, van der Steen AF, van Soest G. Photoacoustic imaging of human coronary atherosclerosis in two spectral bands. *Photoacoustics.* 2014; 2(1):12–20.

33. Jansen K, van der Steen AF, Wu M, et al. Spectroscopic intravascular photoacoustic imaging of lipids in atherosclerosis. *J Biomed Opt.* 2014; 19(2). 026006.

34. Wang B, Karpiouk A, Yeager D, et al. Intravascular photoacoustic imaging of lipid in atherosclerotic plaques in the presence of luminal blood. *Opt Lett.* 2012; 37(7): 1244–1246.

35. Jansen K, Wu M, van der Steen AF, van Soest G. Lipid detection in atherosclerotic human coronaries by spectro scopic intravascular photoacoustic imaging. *Opt Express.* 2013; 21(18):21472–21484.

36. Wang B, Karpiouk A, Yeager D, et al. In vivo intravascular ultrasound-guided photoacoustic imaging of lipid in plaques using an animal model of atherosclerosis. *Ultrasound Med Biol.* 2012; 38(12):2098–2103.

37. Cao Y, Hui J, Kole A, et al. High-sensitivity intravascular photoacoustic imaging of lipid-laden plaque with a collinear catheter design. *Sci Rep.* 2016; 6:25236.

38. Vinegoni C, Botnaru I, Aikawa E, et al. Indocyanine green enables near-infrared fluorescence imaging of lipid-rich, inflamed atherosclerotic plaques. *Sci Transl Med.* 2011; 3(84). 84ra45.

39. Yoo H, Kim JW, Shishkov M, et al. Intra-arterial catheter for simultaneous microstructural and molecular imaging in vivo. *Nat Med.* 2011; 17(12):1680–1684.

40. Bozhko D, Osborn EA, Rosenthal A, et al. Quantitative intravascular biological fluorescence-ultrasound imaging of coronary and peripheral arteries in vivo. *Eur Heart J Cardiovasc Imaging.* 2017; 18(11):1253–1261.

41. Ughi GJ, Wang H, Gerbaud E, et al. Clinical characterization of coronary atherosclerosis with dual-modality OCT and near-infrared autofluorescence imaging. *J Am Coll Cardiol Img.* 2016; 9(11):1304–1314.

42. Verjans JW, Osborn EA, Ughi GJ, et al. Targeted near-infrared fluorescence imaging of atherosclerosis: clinical and intracoronary evaluation of indocyanine green. *J Am Coll Cardiol Img.* 2016; 9(9):1087–1095.

43. Spuentrup E, Botnar RM, Wiethoff AJ, et al. MR imaging of thrombi using EP-2104R, a fibrin-specific contrast agent: initial results in patients. *Eur Radiol.* 2008; 18(9): 1995–2005.

44. Jaffer FA, Vinegoni C, John MC, et al. Real-time catheter molecular sensing of inflammation in proteolytically active atherosclerosis. *Circulation.* 2008; 118(18): 1802–1809.

45. Hara T, Ughi GJ, McCarthy JR, et al. Intravascular fibrin molecular imaging improves the detection of unhealed stents assessed by optical coherence tomography in vivo. *Eur Heart J.* 2017; 38(6):447–455.

46. Htun NM, Chen YC, Lim B, et al. Near-infrared autofluorescence induced by intraplaque hemorrhage and heme degradation as marker for high-risk atherosclerotic plaques. *Nat Commun.* 2017; 8(1):75.

47. Wang H, Gardecki JA, Ughi GJ, Jacques PV, Hamidi E, Tearney GJ. Ex vivo catheter-based imaging of coronary atherosclerosis using multimodality OCT and NIRAF excited at 633 nm. *Biomed Opt Express.* 2015; 6(4): 1363–1375.

48. Kunio M, Gardecki J, Watanabe K, Nishimiya K, Tearney G. TCT-14 assessment of the sources of near-infrared autofluorescence in high-risk coronary artery plaques: a histopathological correlative study. *J Am Coll Cardiol*. 2019; 74(13 Supplement):B14.

49. Bec J, Ma DM, Yankelevich DR, et al. Multispectral fluorescence lifetime imaging system for intravascular diagnostics with ultrasound guidance: in vivo validation in swine arteries. *J Biophotonics*. 2014; 7(5):281–285.

50. Fatakdawala H, Gorpas D, Bishop JW, et al. Fluorescence lifetime imaging combined with conventional intravascular ultrasound for enhanced assessment of atherosclerotic plaques: an ex vivo study in human coronary arteries. *J Cardiovasc Transl Res*. 2015; 8(4):253–263.

51. Bec J, Phipps JE, Gorpas D, et al. In vivo label-free structural and biochemical imaging of coronary arteries using an integrated ultrasound and multispectral fluorescence lifetime catheter system. *Sci Rep*. 2017; 7(1):8960.

52. Lee MW, Song JW, Kang WJ, et al. Comprehensive intravascular imaging of atherosclerotic plaque in vivo using optical coherence tomography and fluorescence lifetime imaging. *Sci Rep*. 2018; 8(1):14561.

# CHAPTER 8

# Quantitative Assessment and Prediction of Coronary Plaque Development Using Serial Intravascular Ultrasound and Virtual Histology

ANDREAS WAHLE[a,b] • LING ZHANG[a,b] • ZHI CHEN[a,b] • HONGHAI ZHANG[a,b] • JOHN J. LOPEZ[c] • TOMAS KOVARNIK[d] • MILAN SONKA[a,b]
[a]Electrical and Computer Engineering, University of Iowa, Iowa City, IA, United States, [b]Iowa Institute for Biomedical Imaging, University of Iowa, Iowa City, IA, United States, [c]Stritch School of Medicine, Loyola University, Maywood, IL, United States, [d]Second Department of Internal Medicine, Charles University, Prague, Czech Republic

## 1. INTRODUCTION

Coronary atherosclerosis is a dynamic and chronic inflammatory process in a vessel wall, leading to the formation of plaques. The natural course of localized plaques can lead to progression (increase of plaque volume), regression (decrease of plaque volume) or they can become stable and do not change for some time.[1] Fast progression of plaque leads to acute coronary syndromes and subsequent major adverse cardiac events (MACE). The identification of patients at risk of such events is challenging but has an enormous impact.[2,3]

In addition to plaque volume, plaque composition plays a crucial role for plaque behavior. Thin-cap fibroatheroma (TCFA) is a precursor for plaque rupture and is the most frequent cause of acute coronary syndromes. Other indicators for high-risk plaque are a large plaque burden and hemodynamically significant reduction in the lumen area ($<4$ mm$^2$). Biology and pathology studies have demonstrated that TCFA is characterized by a large lipid or necrotic core overlying thin fibrous cap ($<65$ μm), which is contributed by macrophages.[4] Virtual histology intravascular ultrasound (VH-IVUS)[5] is an in vivo catheter-based intervention technology and allows the visualization and identification of various plaque phenotypes including TCFA.

VH-IVUS identifies plaque components locally and classifies them as fibro-fatty tissue (FF, light green), fibrous tissue (FT, dark green), necrotic core (NC, red), or dense calcium (DC, white). The identified media is depicted in gray. To classify overall plaque type and severity in a given frame, in agreement with the American Heart Association's Committee on Vascular Lesions[7] definitions, five phenotypes are distinguished once a plaque-burden threshold of 40% is met (Fig. 1)[6]:

- pathological intimal thickening (PIT), mostly containing fibro-fatty plaque components, usually indicating early stages of plaque development;
- fibrous plaque (FP), with fibrous components above a given threshold (15%);
- fibrocalcific plaque (FcP), in addition to fibrous components, containing calcifications above a given threshold (10%);
- thick-cap fibro-atheroma (ThCFA), containing complex fibro atheroma with necrotic core but sufficient fibrous cap thickness ($>65$ μm); and
- thin-cap fibro-atheroma (TCFA), as complex as ThCFA, but with a potentially vulnerable thin fibrous cap ($\leq65$ μm).

Since the fibrous cap can not be directly visualized in IVUS, indirect criteria are used (i.e., NC abutting the lumen over a specific circumferential angle $>30$ degrees and consistently observed over a given number of adjacent frames $\geq3$). While intravascular optical coherence tomography (OCT) can visualize the fibrous cap directly to allow determination of its thickness, it does not provide sufficient penetration to assess the complex plaque structures in depth.

Being able to distinguish which patients would benefit from interventional treatment along with early detection of high-risk lesions is highly desired. Efforts

Intravascular Ultrasound. https://doi.org/10.1016/B978-0-12-818833-0.00008-4

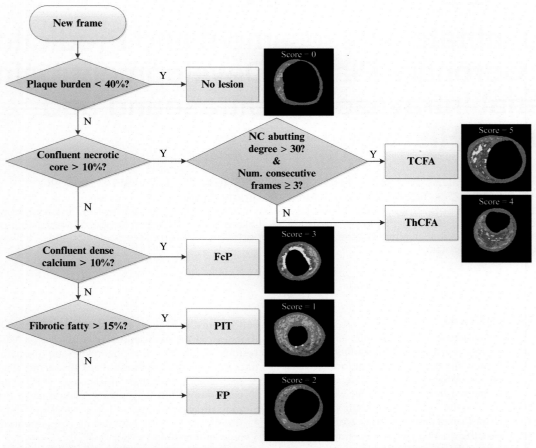

FIG. 1 Hierarchical definition of the VH-IVUS plaque phenotypes. (Reprinted with permission from Zhang L, Wahle A, Chen Z, Lopez JJ, Kovarnik T, Sonka M. Predicting locations of high-risk plaques in coronary arteries in patients receiving statin therapy. *IEEE Trans Med Imaging.* 2018;37(1):151–161. ©2018 IEEE.)

have been undertaken to enable the prediction of treated or untreated outcomes of plaques by correlating local vascular factors to plaque progression and vulnerability.[1,8–12] However, predicting the locations on the coronary artery where TCFA lesions will occur/disappear in the future remains difficult.

We aim for a reliable prediction of plaque that will likely develop to high-risk plaque prone to rupture over a given time. For this purpose, we are utilizing a unique set of VH-IVUS images acquired at baseline and a roughly 12-month follow-up, in an artery that was *not* intervened on. In this way, we can observe the development of plaque over that time and use it to train a classifier, which in turn can predict future plaque development in previously unseen VH-IVUS data at baseline. Fig. 2 outlines the individual steps. Starting with the longitudinal images, the pullbacks are segmented for lumen and adventitia (external elastic membrane) borders, then registered for both location and orientation to obtain a frame-to-frame correspondence

FIG. 2 General pipeline of the IVUS-based prediction system.

between the baseline and follow-up images. From the derived set of features, classifiers are trained to predict individual plaque development. From any given baseline image, with segmentation performed and the required features determined, the classifier can predict on a per-frame basis how the plaque should develop within the trained 12-month time frame.

## 1.1. Segmentation and Registration Approaches

A comprehensive review on the history of well-established IVUS segmentation and quantitative evaluation was presented by Klingensmith et al.[13] Approaches to identify the lumen/plaque and media/adventitia borders include those based on minimum-path graph search,[14] simulated annealing,[15] or active contours.[16,17] Cost or energy functions can be complex and multi-tiered[18] to reflect the nontrivial appearance of IVUS images. In addition, texture information can be taken into account to identify interfaces between regions rather than explicit edges.[19–21] The spatial context between adjacent IVUS frames along the pullback was frequently established, for example, by an initial longitudinal segmentation over the entire pullback followed by cross-sectional contour detection.[22] True 3D methods are preferred but these are computationally more expensive (relative to their respective 2D counterparts). Our approach presented in Section 2 utilizes multidimensional optimal graph search for the simultaneous detection of both surfaces along with advanced editing features based on the LOGISMOS framework.[23]

In order to learn how to prospectively predict high-risk plaques, frame-to-frame registration of baseline and follow-up VH-IVUS pullback data are necessary to facilitate location-specific quantitative comparisons between the two time points. Manual registration of location and orientation is an extremely tedious task and requires expert knowledge. As a result, current clinical baseline/follow-up studies are not utilizing all available information due to undetermined frame location/orientation correspondences. Automated location/orientation registration methods for IVUS pullbacks have attracted substantial research interest in recent years.[24–28] The challenges are associated with motion artifacts, longitudinal oscillations, artifactual angular twisting, variations in the speed of the IVUS catheter during pullback acquisitions, vessel-morphologic changes due to plaque progression/regression at follow-up, and other sources. Fig. 3 provides an example of IVUS pullback registration with seven frame pairs manually registered.

The most intuitive approach to register frame locations is linear fitting (distance normalization) between

two identified landmarks (e.g., side branches).[24,25] This method works well when the transducer pullback exhibits a constant speed, which is not always the case in clinical settings. Nonrigid temporal alignment methods[26–28] (e.g., dynamic time warping [DTW][30]) have the potential to solve this problem. For automated registration of orientation, a cross-correlation-based method (rotating each follow-up image locally[24]) and a Catmull-Rom spline-based method (rotating all the follow-up images globally[25]) have been reported. These two approaches achieve suboptimal rotation results for two already location-aligned IVUS image sequence.

Compared with previous IVUS pullback location/orientation registration methods that treat the registration of location and orientation as separate tasks, Section 3 describes a novel joint spatiotemporal approach for IVUS pullback registration utilizing multidimensional graph search to identify the set of frames with the best correspondence.[29]

## 1.2. Quantitative Assessment and Prediction

Trials such as PROSPECT,[31] VIVA,[32] and ATHEROREMO-IVUS[33] found that high-risk plaques exhibiting TCFA, plaque burden (PB) $\geq$ 70%, and/or minimal luminal area (MLA) $\leq$ 4.0 mm$^2$ are predictors of MACE, as determined by VH-IVUS.[34–36] Although systemic pharmacologic therapies (mostly statins) contribute to regression of coronary plaques,[37,38] high-risk plaques (such as TCFA) still remain.[8,9] Fig. 4 shows in rows (A), (D), (F) plaque regression in the "not high-risk" panel at follow-up, despite the high-risk criteria having been satisfied at baseline; on the other hand, the "high-risk" panel in Fig. 4A, D, F shows examples where the high-risk criteria were also satisfied at follow-up, thus indicating insufficient progression or no change. Additionally, some less-advanced plaques continue to progress to more advanced high-risk plaques even under statin therapy ("high-risk" panel in Fig. 4B, C, E, and G). These residual and newly occurring high-risk plaques remain responsible for MACE in the future. Early identification of locations, in which high-risk plaques will likely develop in the future, is highly desirable as it will enable patient-specific preemptive strategies (such as more intensive pharmacological treatment in high-risk patients or attempting focal plaque stabilization[39]) to avert MACE. Only a very limited number of studies have reported successful prediction of high-risk plaques by showing significant correlation between angiography-IVUS-derived features and TCFA outcome (healed or remaining).[8,9] Clearly, predicting whether, and if so, *where* high-risk plaques exhibiting TCFA, PB $\geq$ 70%, and/or MLA $\leq$ 4 mm$^2$ will remain unchanged or will newly occur remains very challenging.

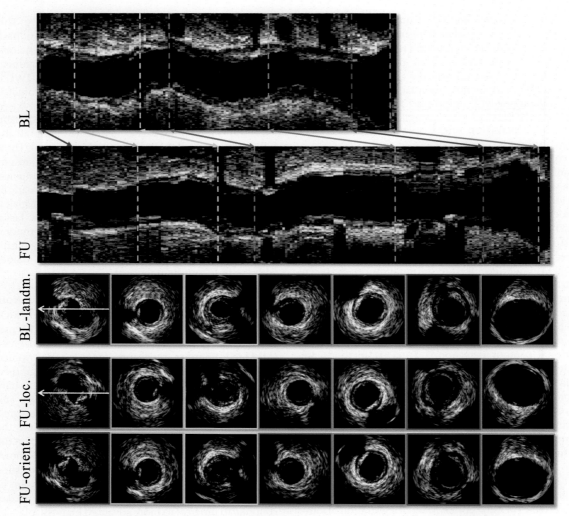

**FIG. 3** Example of IVUS pullback registration. From *top to bottom*: Baseline pullback in longitudinal view; follow-up pullback in longitudinal view; baseline landmark frames in cross-sectional view; corresponding IVUS frames; corresponding follow-up IVUS frames in proper orientation. In this case, the correspondences were determined visually by a human expert. (Reprinted with permission from Zhang L, Wahle A, Chen Z, et al. Simultaneous registration of location and orientation in intravascular ultrasound pullbacks pairs via 3D graph-based optimization. *IEEE Trans Med Imaging*. 2015;34(12):2550–2561. ©2015 IEEE.)

Most previous studies assessed plaque progression in an entire lesion as a whole or over long vessel segments.[1,8,9,11,31–33] Using such long coronary segments fails to reflect the focal nature of clinical events and leads to excessive averaging of focal plaque morphology/composition indices.[12] Historically, prediction of plaque progression relied on statistical modeling, such as logistic regression,[1,12] generating interpretable results but suffering from lower accuracy compared to algorithmic models employing large number of mutually interacting variables.[40,41] Algorithmic modeling, or machine learning,[40,42] allows estimation of prediction accuracy via cross-validation. Machine learning has been effective in coronary plaque component classification,[5] classifying plaque erosion against intact fibrous plaques,[43] and predicting disease recurrence or survival.[41] We have previously demonstrated a potential of machine learning for predicting subsequent development of TCFAs.[44,45]

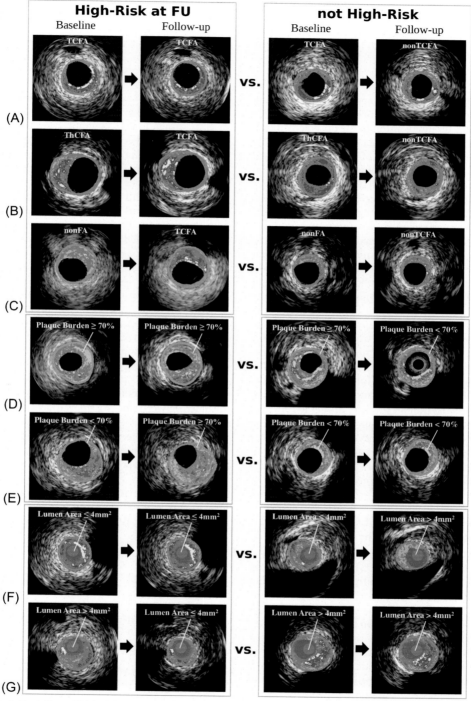

FIG. 4 Evolution of high-risk plaques. Progression, regression, and quiescence of: (A, B, C) TCFA; (D, E) PB ≥ 70%; (F, G) MLA ≤ 4 mm². Plaque tissue compositions are color-coded by virtual histology: dense calcium (DC, *white*), necrotic core (NC, *red*), fibrofatty (FF, *light green*), and fibrotic tissue (FT, *dark green*). *ThCFA*, thick-cap fibroatheroma; *non-FA*, nonfibroatheroma; *FU*, follow-up.

Therefore, we hypothesized that machine learning approaches applied to serial studies of intravascular imaging data and clinical characteristics from patients allow prediction of future high-risk plaque locations and types.

### 1.3. Study Background

The studies presented in this chapter are based on data from 61 patients fulfilling the inclusion criteria (stable angina pectoris, no intervention in the observed artery), selected from a database of 121 patients with stable angina pectoris enrolled in one of the two studies comparing statin therapy for atherosclerosis progression (HEAVEN[46] and PREDICT (NCT01773512), Charles University Hospital, Prague, Czech Republic). Specific to these unique studies are the availability of VH-IVUS image pullback performed in a native coronary artery with angiographically determined maximum stenosis ≤50% at baseline with no indication for either percutaneous coronary intervention (PCI) or coronary artery bypass grafting (CABG), and follow-up imaging 8–14 months after the baseline imaging. In this way, the regression or progression of plaque in these arteries only under the effect of statin therapy could be observed. IVUS imaging was performed using the phased-array probe (Eagle Eye 20 MHz 2.9F monorail, Volcano Corporation, San Diego, CA), IVUS InVision console for the HEAVEN study and s5 console for the PREDICT study, InVision Gold software, and motorized pullback at 0.5 mm/s (research pullback device, model R-100, Volcano Corporation) were used in all acquisitions. The study protocol conforms to the ethical guidelines of the Declaration of Helsinki, was approved by the Institutional Review Board of Charles University, and all patients provided written informed consent.

## 2. GRAPH-BASED IVUS SEGMENTATION
### 2.1. Optimal Graph Search: LOGISMOS

Layered optimal graph image segmentation for multiple objects and surfaces (LOGISMOS) is a general-purpose framework for segmenting multiple $n$-D surfaces that mutually interact within individual and/or between objects.[47–50] Columns of interconnected graph nodes are used to cover the search region for the target surfaces. A cost function is used to assign each graph node with a cost value that represents the unlikeliness of the target surface passing through the node. Such a graph construction process converts the problem of finding the multisurface segmentation to finding the set of nodes, one per column, with globally minimal total cost. In addition, prior shape and anatomy knowledge can also

be incorporated into the graph as context-specific graph arcs that enforce geometric constraints. After assigning each graph node with unlikeliness-based cost, the set of nodes (one per column) with minimal total cost can be located by finding the minimum $s$–$t$ cut of the graph.

### 2.2. Graph and Cost-Function Design

For IVUS pullbacks, to segment the tubular target surfaces as shown in Fig. 5a, graph columns starting from the cross-sectional centers of the IVUS frames are generated, with a fixed number of columns per cross-sectional slice and a constant polar angle increment. As shown in Fig. 5b, graph nodes belonging to the same column are first connected with (here: vertical) intracolumns arcs. Surface smoothness constraints, $\Delta_s$, are incorporated into the graph as intercolumn arcs. The constraint $\Delta_s$ guarantees that if the segmentation identifies the $i$th and $j$th nodes from two neighboring columns to be the on-surface nodes, then $|i - j| \leq \Delta_s$. Note that for IVUS images, it is possible to use two different $\Delta_s$ values, one for columns within the same cross-sectional slices, and another for columns from neighboring cross-sectional slices. For an object with multiple surfaces that share a common topology, each individual surface is associated with a subgraph constructed similar to that in Fig. 5b. The relationship between a pair of surfaces can be controlled by a set of surface separation constraints, $\Delta_l$ and $\Delta_u$, which represent the lower and upper bounds for the distance between noncrossing surfaces. The surface separation constraints are incorporated into the graph as arcs that are connecting corresponding columns from the two subgraphs as shown in Fig. 5c.

Both graph topology and constraints, as well as the cost function employed, substantially influence the segmentation results. The cost function initially developed in Ref. 18 was optimized for Boston Scientific 40 MHz IVUS catheters and was modified to

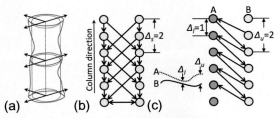

FIG. 5 LOGISMOS graph construction for IVUS segmentation. (a) Establishing columns for tubular target surface. (b) Adding graph arcs for single-surface segmentation. (c) Adding arcs for multisurface segmentation with surface separation constraints.

accommodate the different appearance of the Volcano EagleEye 20 MHz catheters. As a cost function for outer boundary (adventitia surface), gradient-based edge costs are utilized. As for the inner boundary (luminal surface), edge costs are combined with Rayleigh probability density inspired in-region costs, which are well suited to describe the typical speckle noise pattern found in ultrasound images.[23]

## 2.3. Advanced Editing: Just Enough Interaction

The minimum $s$–$t$ cut problem for LOGISMOS segmentation is solved by focusing on the dual problem of finding the maximum flow, for which many algorithms are available. The excesses incremental breadth-first search (EIBFS) algorithm achieves the goal by iteratively pushing more flow into the network of nodes and arcs until it saturates, and also allows applying changes to a saturated network.[51] When such changes are only occurring on a small portion of the network, EIBFS can achieve a new solution within a very small fraction of the time that it took to achieve the previous solution.

The just-enough interaction (JEI) extension of LOGISMOS utilizes the above dynamic nature of the EIBFS algorithm to edit the segmentation result via interactive modification of cost while maintaining the global optimality and geometric constraints of the segmentation. The user simply draws 2D contours close to the parts requiring editing by marking a few points, thus indicating the correct locations of the surface. This should be considered a "hint" to the LOGISMOS-JEI algorithm rather than manually redrawing a contour. The graph nodes close to these contours will be assigned to lower costs to find the new solution. The new segmentation can often be achieved near real time, and the 2D interaction often produces 3D surface changes that eliminate the necessity of 3D slice-by-slice retracing.

The initial implementation and user interface presented in Ref. 23 has been replaced with a flexible modular framework using a Qt5 frontend and can be easily applied/customized to various cardiac and noncardiac applications.[52] Specifically, this framework is better suitable for the physicians to use directly and in time-efficient manner. The only step that requires user interaction is the IVUS segmentation refinement, which requires just 6 min of expert time on average.

## 3. BASELINE/FOLLOW-UP REGISTRATION

Our system registers image sequences in a highly automated manner as long as sufficiently accurate lumen and adventitia segmentations are available for all frames of both image sequences. This means that there is no need to provide a variety of manually obtained guiding information, such as branch locations/orientations,[24–27] plaque characterization,[24,26,27] matching anatomic landmarks,[24,25] and coregistration angles.[24] Compared with previous IVUS pullback location/orientation registration methods that treat the registration of location and orientation as separate tasks, we have developed a novel joint spatiotemporal approach for IVUS pullback registration.[29] By combining these two aspects in one global optimization task of finding an optimal path in a 3D graph, the location and orientation are registered simultaneously. To increase the robustness to the changes of vessel morphologic features, our method combines advantages from feature-based and direct approaches by incorporating plaque thickness and plaque/perivascular pixel similarities, which provides more robust correspondence in two time points. To ensure geometrically feasible registration, graph arcs incorporate prior information about baseline/follow-up correspondences of the two IVUS sequences as well as information about limited-range angular twisting between consecutive IVUS image frames. To initialize the registration procedure, global and local similarities of two IVUS pullbacks are extracted from the 3D graph to automatically identify the most proximal and most distal image pair correspondences. To the extent of our knowledge, our approach is the first in the literature to simultaneously establish temporal (location) and spatial (orientation) correspondences between two different image sequences of two similar dynamic scenes by using a graph-based method.

## 3.1. 3D Correspondence Graph

Given a pair of baseline and follow-up IVUS pullbacks, a 3D directed graph is constructed and searched for the optimal path (Fig. 6). In the 3D graph, each node represents the possible location and orientation correspondences of a baseline and a follow-up image frame pair. For example, the uppermost green graph node with coordinates ($bl = 4$, $\theta = 212$ degrees, $fu = 3$) in Fig. 6 represents correspondence of a follow-up frame no. 3 with 212 degrees rotation with a baseline frame no. 4. The complete registration is defined by a sequence of nodes, which for each ($bl$, $\theta$, $fu$) coordinate triplet forms a path in the 3D graph defining a possible location and orientation correspondence of baseline and follow-up image frames. Costs assigned to each node in the 3D graph reflect similarities of IVUS image data appearance and plaque morphology between the baseline and follow-up. An optimal path is defined as the path with

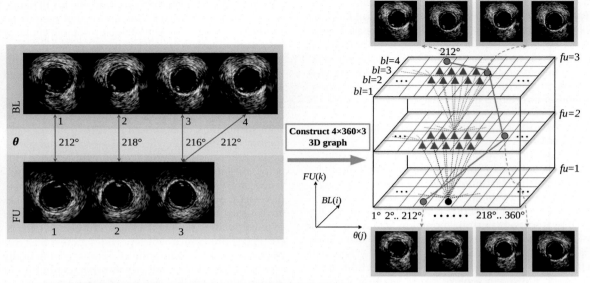

**FIG. 6** 3D graph used for solving the registration problem. This illustrative example assumes a baseline pullback (four frames) and a follow-up pullback (three frames). The possible connections for a path from the black-node example at ($bl = 1$, $\theta = 214$ degrees, $fu = 1$) with an assumed $\theta_{twist} = \pm 2$ degrees are shown. Green nodes and connections depict the optimal path through the graph, determining the registration solution.

the lowest global cost, which in an overall sense defines the frame-to-frame location and orientation registration of two image sequences. To calculate the lowest cost path, a dynamic programming algorithm is utilized.[53,54] A three-dimensional cumulative cost matrix $C$ is computed according to node costs $c(i, j, k)$ at ($bl = i$, $\theta = j$ degrees, $fu = k$).

The design of the node connections assumes limited angular twisting between consecutive image pairs.[24,55] In each stage of the dynamic programming, the best one of the $N = 3 \times (2 \times \theta_{twist} + 1)$ possible preceding nodes is selected, where $\theta_{twist}$ is the rotational constraint of angular frame-to-frame twisting. Another constraint is the inability to reverse the direction in one of the pullbacks, thus eliminating the ($bl = i - 1$, *, *) and (*, *, $fu = k - 1$) nodes as possible successors to the nodes in ($bl = i$, *, *) and (*, *, $fu = k$). A frame in one pullback may be assigned to multiple successive frames in the other pullback (e.g., due to different heart rates). An example for the possible node connections for a single starting node (black) is visualized in Fig. 6.

## 3.2. Feature- and Morphology-Based Similarity

The design of the node-associated costs is always problem-specific. For IVUS pullback registration, each

cost $c(i, j, k)$ reflects the similarity between the $i$th baseline image and the $k$th follow-up image with $j$-degree rotation. The similarity is designed as a combination of feature-based and direct approaches based on three terms: (1) perivascular tissue, which includes pixels outside of the adventitia border; (2) plaque appearance, the pixels between the adventitia and lumen borders; and (3) plaque thickness, defined as the distance between lumen and adventitia borders at 360 circumferential wedges centered at the lumen centroid. This cost function design follows the rules for characteristic calcifications, perivascular landmarks, and plaque shape, which cardiologists use to match baseline and follow-up IVUS pullbacks.[56] For each term, the noncorrelation $1 - Corr(i, j, k)$ is calculated based on the normalized cross-correlation approach detailed by Timmins et al.[24] The node costs $c(i, j, k)$ are the weighted sum of the individual noncorrelations for the three terms, where a size factor was added to the plaque-thickness term (3) to discriminate between two similar plaque shapes with different sizes.[29] Note that when computing these $Corr()$ values, coinciding lumen centroids for the baseline and rotated follow-up image pairs are provided by our automated segmentation and operator-guided refinement algorithm.[23]

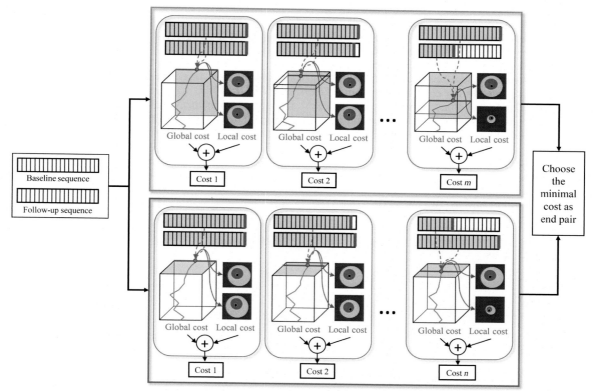

FIG. 7 Proposed global and local identification of *ending* pair. Given a baseline and follow-up IVUS sequence pair, our method constructs a 3D graph and follows two tracks: in the first track, the last baseline frame is fixed and the follow-up frame iteratively shifted by the minimal required length of baseline/follow-up overlap (25 mm in our case). In each iteration, global and local costs embedded in the 3D graph are extracted and combined. By fixing the last follow-up frame and iteratively shifting the baseline frame, another series of costs are obtained in the second track. The frame pair providing minimal cost is chosen as the *ending* pair. (Reprinted with permission from Zhang L, Wahle A, Chen Z, et al. Simultaneous registration of location and orientation in intravascular ultrasound pullbacks pairs via 3D graph-based optimization. *IEEE Trans Med Imaging*. 2015;34(12):2550–2561. ©2015 IEEE.)

## 3.3. Finding the Start and End Frame Pairs

For the baseline and follow-up IVUS acquisition in the same vessel, the *starting* and *ending* positions of pullbacks are not always the same due to operator- or equipment-related variations in the imaging process. Therefore, before applying the proposed registration method, we need to identify the most proximal and most distal corresponding image frame pairs. Our strategy combines global and local costs embedded in the 3D graph-based framework. Fig. 7 demonstrates identification of the *ending* frame pair. The *global* cost corresponds to the mean cost of the optimal path and thus represents the average similarity between two registered IVUS image sequences; the *local* cost is associated with the similarity between the $i$th image in baseline and the $k$th image with rotation angle $\theta_j$ in

follow-up. After the *ending* pair is identified, both the 3D cumulative cost matrix $C$ and the cost matrix $c$ are inverted and the same global and local approach is used to identify the *starting* pair with a searching range constrained by the minimal required lengths of the baseline/follow-up pullback overlap (25 mm overlap required in our studies). Subsequently, the entire selection process is reversed by identifying the *starting* pair first followed by obtaining the *ending* pair. Thus, two start/end pairs are available and the final solution is identified by choosing the lower total cost (global and local) from these two solutions. The 3D graph is subsequently constructed and the optimal path is determined, simultaneously registering the IVUS pullback with respect to location and orientation.

### 3.4. Validation Results

Fig. 8 shows an example for the registration, along with the path through the 3D graph. Summarizing the validation results on a subset of 29 patients in Ref. 29, our 3D graph-based method achieved mean distance errors (relative to three expert registrations) ranging from 0.72 to 0.79 mm with the mean angle errors ranging from 7.3 to 9.3 degrees. There were no significant differences between our method and experts in location and orientation registration ($P = \text{NS}$). The maximal distance and angle errors generated by our method were 8.4 mm and 147 degrees, respectively, comparable to the maximum disagreements between the experts. Evaluation of 383 landmarks on the full set of 61 patients demonstrated an average registration error of $0.45 \pm 0.78$ mm. The registration produced 6341 registered IVUS frame pairs (46–191 frame pairs per patient) with 0.23–0.70 mm frame-to-frame separation distances between adjacent frames depending on the heart rate.

## 4. MORPHOLOGY ASSESSMENT: REMODELING

As part of the predictive investigation of serial development of human coronary atherosclerosis, we were investigating the effect of coronary remodeling as described by Glagov et al.[57] Arterial remodeling consists mainly of two types of tissue transformation—plaque accumulation and adventitia deformation, which together have been identified as the determinant of lumen size and may equally contribute to the luminal narrowing.[58]

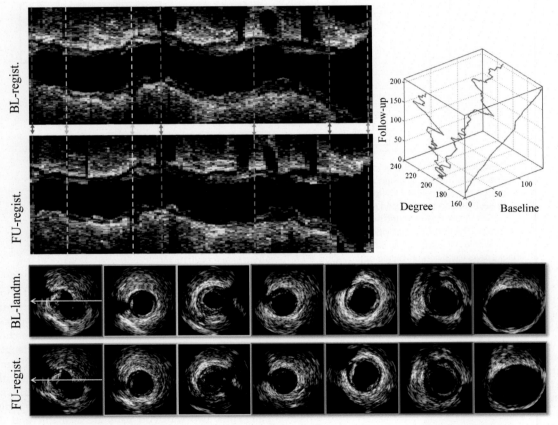

**FIG. 8** Automated registration of pullback pair from Fig. 3. From the *first to the fourth row*: registered baseline pullback; registered follow-up pullback; baseline landmarks selected by Expert 1; registration results obtained by our automated method. Optimal path (*green path*) in 3D (*bl, θ, fu*) space and its projections (*red paths*) on 2D (*bl, fu*) plane and 2D (*θ, fu*) plane generated by our method for this example are shown in the *upper-right panel*. (Reprinted with permission from Zhang L, Wahle A, Chen Z, et al. Simultaneous registration of location and orientation in intravascular ultrasound pullbacks pairs via 3D graph-based optimization. *IEEE Trans Med Imaging*. 2015;34(12):2550–2561. ©2015 IEEE.)

Earlier IVUS studies, such as Refs. 59–62, suffered from the lack of serial measurements over time and/or the ability to analyze all available images of the vessel. Thus, they were forced to use an indirect remodeling index as a ratio of cross-sectional vascular area relative to a local reference area coming from the same vessel. This process is observer dependent and may not result in a sufficiently plausible selection of the reference location in the presence of diffuse atherosclerosis. In contrast, our study can directly observe the remodeling process as the difference in plaque and adventitia cross-sections between the two time points, at baseline versus follow-up imaging. The following description is an extension of our initial report based on 31 patients,[63] now using the full set of 61 patients, in essence confirming the findings published earlier.

## 4.1. Progression and Regression Groups

Since 48 out of the 61 patients received statin lipid-lowering therapy before enrollment and all of them received statin treatment after enrollment, the observed remodeling process is under the influence of treatment and as such does not necessarily reflect the natural development of plaque and adventitia. Consequently, the majority of patients experienced a regression of plaque. Zarins et al.[64] proved that different segments of the same artery may respond differently to increasing atherosclerotic plaque, thus warranting a per-frame analysis of the remodeling process in addition to or as a replacement for the patient-wise more global assessment.

In total, 5215 cross-sectional IVUS-VH frame pairs were available for this analysis. To better assess plaque CSA magnitude changes, we divided the frames into two groups, plaque regression ($n = 3058$) and plaque progression ($n = 2157$), based on plaque CSA developing direction. On the patient level, plaque regression (progression) percentage for each patient was defined as the number of frames experiencing plaque CSA decrease (increase) at follow-up divided by the length of the whole pullback.

## 4.2. Plaque and Adventitia Development

The majority of all frames (58.6%) experienced decreased plaque cross-sectional area (CSA) from baseline to follow-up, while 41.4% of frames experienced the opposite during statin treatment. According to the frequency distribution of plaque CSA changes presented in Table 1, plaque regression (baseline plaque CSA > follow-up plaque CSA) is more likely to happen during lipid-lowering therapy. Only 10.4% of the frames showed a plaque CSA decrease or increase in excess of 2 mm². The patient-level plaque regression percentage for all 61 patients was 58.2% ± 25.1%, thus the mean

### TABLE 1
### Frequency Distribution of Plaque Cross-Sectional Area Changes

| mm² | n | % |
|---|---|---|
| Decreased plaque CSA | | |
| <0 | 3058 | 58.6 |
| <−1 | 1129 | 21.6 |
| <−2 | 407 | 7.8 |
| <−3 | 181 | 3.5 |
| <−4 | 92 | 1.8 |
| <−5 | 43 | 0.8 |
| Increased plaque CSA | | |
| >0 | 2157 | 41.4 |
| >1 | 579 | 11.1 |
| >2 | 136 | 2.6 |
| >3 | 39 | 0.7 |
| >4 | 14 | 0.3 |
| >5 | 3 | 0.1 |
| | (Total frames, $n = 5215$) | |

being close to the per-frame distribution, however, showing high interpatient variability.

Analyzing adventitia development, 61.1% of all frames underwent decreasing adventitia CSA from baseline to follow-up while 38.9% underwent increasing CSA. The directional relationship between adventitia and plaque CSA development is presented in Table 2. It appeared that majority of frames (48.2% and 28.5%) experienced adventitia CSA changes in the "same" direction as plaque CSA changes ($P \ll .001$). Magnitude development relationship between adventitia and plaque CSA was investigated by linear mixed regression.

### TABLE 2
### Directional Relationship Adventitia Versus Plaque Area Changes

| | Adventitia regression, n (%) | Adventitia progression, n (%) |
|---|---|---|
| Plaque regression | 2515 (48.2) | 543 (10.4) |
| Plaque progression | 670 (12.8) | 1487 (28.5) |
| Chi-squared test | $P \ll .001$ | |

If plaque and adventitia CSA developed in the *same* direction:

- Unsigned adventitia CSA regression magnitude (8.1% ± 8.6%) was positively associated with unsigned plaque CSA regression magnitude ($\beta = 0.21$, $P \ll .001$).
- Adventitia CSA progression magnitude (9.6% ± 12.0%) was positively associated with plaque CSA progression magnitude ($\beta = 0.29$, $P \ll .001$).

If plaque and adventitia CSA developed in the *opposite* direction:

- Unsigned adventitia CSA regression magnitude (3.9% ± 3.5%) was slightly associated with plaque CSA progression magnitude ($\beta = 0.02$, $P = .01$).
- Adventitia CSA progression magnitude (5.1% ± 7.4%) was not associated with plaque CSA regression magnitude ($P = .17$).

On the patient level, adventitia regression percentage for each patient was defined as the number of frames experiencing adventitia CSA decrease at follow-up divided by the length of whole pullback. Adventitia regression percentage and plaque regression percentage for all patients were positively correlated ($R = 0.76$, $P \ll .001$). Adventitia regression (progression) magnitude for each patient was defined as the average of frame-based adventitia regression (progression) magnitude per patient. Adventitia and plaque development magnitude for all patients were positively correlated.

## 5. PREDICTION OF PLAQUE DEVELOPMENT

The overall design of training process and application of the predictive classifier is shown in Fig. 9. Baseline and follow-up VH-IVUS pullback data are first segmented and mutually registered. The location-specific (frame-level) features and systemic/demographic information at baseline undergo feature selection[65,66] to form an optimal feature subset. A set of support vector machine (SVM) classifiers[67,68] is trained to predict focal plaque type at follow-up. To predict three high-risk plaque forms (TCFA, PB ≥ 70%, MLA ≤ 4 mm²) at follow-up, seven binary classifiers are trained according to the transitions shown in Fig. 4:

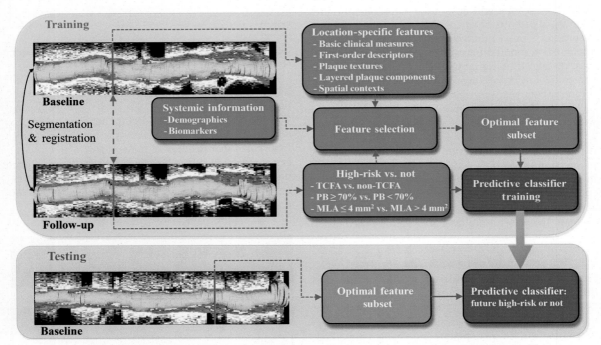

FIG. 9 Prediction of future high-risk plaque locations. Lumen shown in *orange*, plaque tissue composition color-coded by VH. Seven separate classifiers are trained to predict high-risk plaques. (Reprinted with permission from Zhang L, Wahle A, Chen Z, Lopez JJ, Kovarnik T, Sonka M. Predicting locations of high-risk plaques in coronary arteries in patients receiving statin therapy. *IEEE Trans Med Imaging.* 2018;37(1):151–161. ©2018 IEEE.)

- Three classifiers predict whether focal plaque types (1) TCFA, (2) thick-cap fibroatheroma (ThCFA), and (3) nonfibroatheroma (non-FA) at baseline transition to TCFA plaque type or not at follow-up (Fig. 4A–C).
- Two classifiers predict whether local plaques with (4) PB $\geq$ 70% or (5) PB < 70% at baseline lead to follow-up plaque burden PB $\geq$ 70% or not (Fig. 4D and E).
- For plaques with (6) MLA $\leq$ 4 mm$^2$ or (7) 4 mm$^2$ < MLA $\leq$ 6 mm$^2$ at baseline, two classifiers predict whether MLA $\leq$ 4 mm$^2$ or not at follow-up (Fig. 4F and G).

In all cases, training and testing sets were disjoint in a leave-one-patient-out (LOPO) cross-validation manner. Each of the 61 training/validation sessions had a different patient "left out" resulting in a complete separation of training and validation at the patient level and allowing validation on a sufficiently large set. The classification success was therefore assessed as average performance of these 61 independently trained classifiers.

## 5.1. Plaque Phenotype Definitions

The American Heart Association's Committee on Vascular Lesions[7] has classified the criteria for plaque phenotypes into six categories (Fig. 1): no lesion (NL), pathological intimal thickening (PIT), fibrous plaque (FP), fibrocalcific plaque (FcP), ThCFA, and TCFA. In Refs. 6, 69, we assigned a vulnerability score in the range 0–5 for each phenotype (NL = 0, PIT = 1, FP = 2, FcP = 3, ThCFA = 4, and TCFA = 5). Considering that a large NC component and existence of a thin fibrous cap <65 µm indicate instability of plaque,[70] three NC- and NC-lumen-related measurements were introduced, that is, percentage of the maximal confluent NC region, angle, and size (number of pixels) of NC-region abutting the vessel lumen. The TCFA category depends on confluent NC > 10% and NC abutting the lumen >30 degrees in at least three consecutive frames, as was done in the PROSPECT study.[31]

## 5.2. Features and Classifiers

Two categories of features including location-specific features ($m = 236$) and systemic information ($m = 18$, demographic information and biomarkers)[1,31,69] were extracted to be used for training the prediction classifier. The location-specific features further contain four subtypes including basic clinical measurements ($m = 21$), first-order descriptors ($m = 9$), plaque textures ($m = 16$), layered plaque components ($m = 72$), and spatial contextual features ($m = 118$). Among them, basic clinical measurements reflect local vascular disease severity and/or plaque development,[31,34,36,71] others were

inspired by feature descriptors successfully used in computer-aided diagnosis applications (first-order descriptors, plaque textures)[72,73] or incorporated spatial information in circumferential and axial directions (layered plaque components, spatial contextual features).

In addition to the phenotype score, the percentages of each plaque component DC/NC/FF/FT along with the derived NC-related indices, the CSAs (lumen and adventitia), and derived indices (plaque thickness, plaque burden, eccentricity, and remodeling index[71]) were determined for each frame. The frame distance from the ostium was derived from the reconstructed 3D vessel model[55] and obtained for all frame locations. To enable a more detailed examination of plaque components, a layered analysis of different plaque depths in circumferential direction was performed in wall rings, separating an "inner" ring from an "outer" ring using adaptive radii of 10%, 20%, …, 90% distance between the lumen and adventitia borders. Percentages of DC/NC/FF/FT in the 10%–90% inner and outer rings were calculated, and therefore 72 layered plaque component features were obtained for each plaque. Fig. 10 shows examples of analysis of layered plaque components in a remaining TCFA and a healed TCFA. Further features include first-order gray-level/intensity-based indices, as well as $4 \times 4$ cooccurrence matrices[74] (VHLCMs) for each frame, based on the four-color-coded DC/NC/FF/FT map and four directions ($\theta = 0, 45, 90, 135$ degrees). Second-order statistical texture features including contrast, correlation, energy, and homogeneity were calculated from each VHLCM.

To incorporate spatial context between adjacent locations in axial direction and also to limit the impact of noise (e.g., potentially inaccurate frame-to-frame registration), we further extended the feature sets by computing the average values of features in the adjacent distal and proximal frames, resulting in 118 spatial contextual features. The complete set of features used in this study is presented in Ref. 6.

## 5.3. Feature Selection for Predictions

Feature selection was performed to increase the predictor performance and provide better understanding of the underlying factors/features that contribute to development of high-risk plaque. Fig. 11 illustrates the feature selection and predictive model validation procedure. The features from the complete feature set ($M = 254$) were ordered by a support vector machine recursive feature elimination (SVM RFE) method according to the discrimination ability of feature subset.[65] SVM RFE is an application of RFE using the weight magnitudes of features in SVM[67,68] training as the

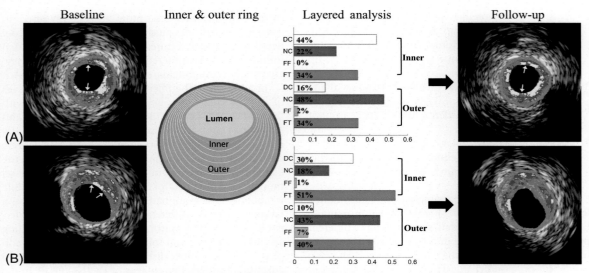

**FIG. 10** Layered plaque components. Baseline VH-IVUS images (*first column*), inner (*green*) and outer (*orange*) ring (*second column*), layered plaque components (*third column*), and follow-up VH-IVUS images (*fourth column*). (A) TCFA unchanged. (B) TCFA healed (regressed to ThCFA, 20% ring used). (Reprinted with permission from Zhang L, Wahle A, Chen Z, Lopez JJ, Kovarnik T, Sonka M. Predicting locations of high-risk plaques in coronary arteries in patients receiving statin therapy. *IEEE Trans Med Imaging*. 2018;37(1):151–161. ©2018 IEEE.)

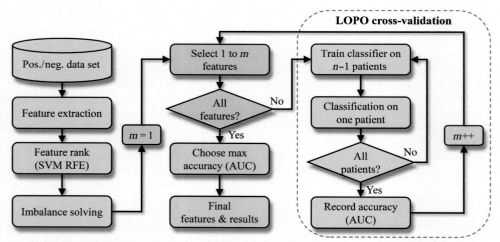

**FIG. 11** Feature selection and predictive model validation. (Reprinted with permission from Zhang L, Wahle A, Chen Z, Lopez JJ, Kovarnik T, Sonka M. Predicting locations of high-risk plaques in coronary arteries in patients receiving statin therapy. *IEEE Trans Med Imaging*. 2018;37(1):151–161. ©2018 IEEE.)

ranking criterion, and is very effective for discovering/ranking informative features while overfitting is prevented.[65,72] Each feature was first normalized, then linear SVMs (parameter $C = 1$) were trained followed by the iterative chunk-based elimination approach[65] to produce a final feature subset of 50 features. Because

of the imbalance in the class labels in our study (e.g., 55 TCFA remained while 353 TCFA healed), a different error costs (DEC) method[75] was utilized in SVM training, in which the individual class weights in the SVM model were assigned to be inversely proportional to the imbalance ratio.

An incremental LOPO cross-validation with incrementally increasing number of features was repeatedly performed until all 50 best-ranked features were considered, as shown on the right-hand side in Fig. 11. As detailed in Ref. 6, various imbalanced data learning approaches were examined. The prediction performance including sensitivity (SEN), specificity (SPE), and the area under a receiver operating characteristic curve (AUC) were calculated and recorded after testing all $n$ patients. In addition, as is widely accepted in class imbalance learning literature,[76] G-mean metric (G-mean = $\sqrt{sensitivity \times specificity}$) was used to quantify the prediction accuracy. Finally, the feature subset with the highest AUC was chosen as the final feature set, and the final prediction rates were obtained.

The prediction strategy was designed to predict the future type of plaque for each IVUS frame location. Considering that the TCFA definition usually requires presence of TCFA in at least three consecutive frames, a postprocessing step adjusts the local TCFA type labeling accordingly.[45] To take the current clinical practice scenario into account, we generalize our frame-level prediction to lesion and patient levels. A lesion was defined as at least three consecutive IVUS frames with PB $\geq$ 40%,[31] and lesions were considered separate if there was $\geq$5 mm segment with PB < 40% between them.[9] A lesion-/patient-level prediction would be labeled as TCFA, PB $\geq$ 70%, or MLA $\leq$ 4 mm$^2$ only if at least three consecutive frames in a lesion/patient were predicted to be labeled as TCFA, PB $\geq$ 70%, or MLA $\leq$ 4 mm$^2$. Fig. 12 shows an example of follow-up high-risk plaque locations as predicted from the baseline pullback.

## 5.4. Analysis and Prediction Results

At baseline, the 61 VH-IVUS pullbacks from the 61 enrolled patients exhibited 408 TCFA, 1068 ThCFA, and 4865 non-FA locations, as measured per image frame. At follow-up, 55 (13.5%) TCFAs remained TCFAs, 36 (3.4%) ThCFAs, and 33 (0.7%) non-FAs progressed to TCFAs. During this period, 353 (86.5%) TCFAs healed (transformed to non-TCFA), while 1032 (96.6%) ThCFAs and 4832 (99.3%) non-FAs did not progress to TCFA at follow-up (Table 3). There were 206 (3.2%) locations of PB $\geq$ 70% at baseline. The total number of frames with PB $\geq$ 70% at follow-up decreased to 173 (2.7%): 112 locations (64.7%) with PB $\geq$ 70% at baseline and 61 frames (35.3%) with PB < 70%, despite lipid-lowering therapy. Frames with baseline PB < 50% never progressed to PB $\geq$ 70% at follow-up. There were 303 (4.8%) frames with MLA $\leq$ 4 mm$^2$ at baseline, decreasing to 267 frames (4.2%) at follow-up: 177 frames (66.3%) maintained

MLA $\leq$ 4 mm$^2$ and 90 frames (33.7%) progressed to MLA $\leq$ 4 mm$^2$. Frames with baseline MLA > 6 mm$^2$ never regressed to MLA $\leq$ 4 mm$^2$ at follow-up.

Out of the 162 features involved in all 7 predictors, 9.3% reflect basic clinical measures, 8.6% first-order descriptors, 11.7% layered plaque components, 7.4% plaque textures, 42.6% spatial contexts, and 20.4% systemic/demographic information. Based on the LOPO cross-validation results, our models predict (1) TCFA, (2) ThCFA, and (3) non-FA plaque types at follow-up with high G-mean values of 85.9%, 81.7%, and 77.0%, respectively (sensitivities and specificities ranging from 63.6% to 93.3%). The plaque burden categories (1) PB $\geq$ 70% and (2) 50% $\leq$ PB < 70% were predicted with high correctness (G-mean) of 80.8% and 85.6%, respectively (sensitivities and specificities ranging from 77.7% to 85.9%). Finally, the lumen area (1) MLA $\leq$ 4 mm$^2$ and (2) 4 mm$^2$ < MLA $\leq$ 6 mm$^2$ at follow-up were predicted with high G-mean values of 81.6% and 80.2%, respectively (sensitivities and specificities ranging from 77.0% to 86.4%). Table 4 presents the accuracy of the predictions on the lesion and patient levels.

Our study predicts future MACE-related plaques phenotypes (TCFA, PB $\geq$ 70%, MLA $\leq$ 4 mm$^2$) with high sensitivity and specificity, unlike previous studies,[1,8,9,11,12] which focused on relationships between baseline predictors and plaque progression/regression. The proposed approach achieved a more precise and frame-location specific prediction ($\approx$0.5 mm) compared to the segment level ($\approx$3 mm)[1] or lesion level ($\approx$10 mm)[8,9] locational accuracy. While we have developed and evaluated our predictive models using data from patients with stable angina, our strategy is not limited to this patient group and can be easily applied (trained) on further patient cohorts.

## 6. CONCLUSIONS

The prediction of future plaque development, and especially the early identification of plaque that will likely develop into high-risk plaque and thus needs attention, is of utmost importance for targeted and patient-specific therapy planning and the prevention of future major adverse cardiac events.

Within the scope of this study, we have developed a pipeline of individual components for segmenting entire IVUS pullbacks, registering pullbacks from multiple time points on a frame-by-frame basis, determination and quantification of a complex set of features, and multiple predictors to identify future high-risk

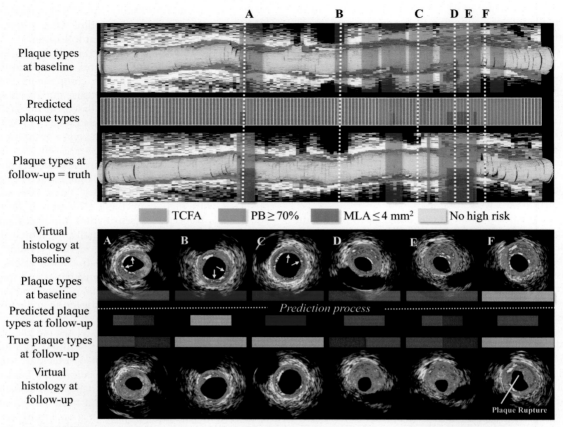

**FIG. 12** Future high-risk plaque locations. *Upper panel*: Baseline VH-IVUS pullback with high-risk plaque locations (*first row*), predicted follow-up high-risk plaque locations (*second row*), and ground truth at follow-up (*third row*). TCFA in *red*; PB $\geq$ 70% *blue*; MLA $\leq$ 4 mm$^2$ *purple*. *Lower panel*: Cross-sectional VH-IVUS frames at the six locations marked in the *upper panel*. (A) TCFA healed, PB $\geq$ 70% unchanged, and new MLA $\leq$ 4 mm$^2$ occurred. (B) TCFA healed. (C) TCFA healed. (D) PB $\geq$ 70% and MLA $\leq$ 4 mm$^2$ unchanged. (E) PB $\geq$ 70% unchanged while new MLA $\leq$ 4 mm$^2$ present. (F) No high-risk plaque with plaque rupture at follow-up. (Reprinted with permission from Zhang L, Wahle A, Chen Z, Lopez JJ, Kovarnik T, Sonka M. Predicting locations of high-risk plaques in coronary arteries in patients receiving statin therapy. *IEEE Trans Med Imaging*. 2018;37 (1):151–161. ©2018 IEEE.)

plaques just from baseline imaging and demographic/biomarker data. Our machine-learning approach demonstrated that location-specific prediction of future plaque phenotypes related to MACE is feasible, thus improving risk stratification in patients with established coronary artery disease. The newly designed VH-IVUS features improve the prediction accuracy compared to the conventional approach of only employing clinical measures. Our method is general, as it can predict three different plaque-type characteristics by employing the same machine-learning approach for all of these three tasks. This characteristic makes our approach easily extendible to other plaque-development prediction tasks, such as prediction of longitudinal remodeling patterns or—once larger data-sets become available—prediction of cardiovascular events.

Once the predictors are trained, the actual analysis task is reduced to segmenting previously unseen single-time point IVUS data and then running the prediction (no registration required). Using the given framework, this is a fast and efficient way to determine likely future

**TABLE 3**
**Plaque Phenotype Changes (Frame Level)**

| Baseline | | FOLLOW-UP | | | | | |
|---|---|---|---|---|---|---|---|
| | | TCFA | ThCFA | FcP | FP | PIT | NL |
| TCFA | 408 (6%) | 55 | 148 | 13 | 109 | 71 | 12 |
| ThCFA | 1068 (17%) | 36 | 390 | 59 | 205 | 277 | 101 |
| FcP | 140 (2%) | 2 | 45 | 32 | 12 | 35 | 14 |
| FP | 826 (13%) | 4 | 108 | 19 | 314 | 279 | 102 |
| PIT | 2005 (32%) | 17 | 128 | 20 | 719 | 881 | 240 |
| NL | 1894 (30%) | 10 | 51 | 18 | 86 | 196 | 1533 |
| Total | 6341 | 124 (2%) | 870 (14%) | 161 (3%) | 1445 (23%) | 1739 (27%) | 2002 (32%) |

Notes: Values are n or n (%). Plaque phenotypes were automatically labeled as given in Ref. 69.
*TCFA*, virtual-histology derived thin-cap fibroatheroma; *ThCFA*, thick-cap fibroatheroma; *FcP*, fibrocalcific plaque; *FP*, fibrous plaque; *PIT*, pathological intimal thickening; *NL*, no lesion.
Source: Reprinted with permission from Zhang L, Wahle A, Chen Z, Lopez JJ, Kovarnik T, Sonka M. Predicting locations of high-risk plaques in coronary arteries in patients receiving statin therapy. *IEEE Trans Med Imaging*. 2018;37(1):151–161. ©2018 IEEE.

**TABLE 4**
**High-Risk Plaque Prediction Results, Lesion, and Patient Levels**

| | No. of TP/P | No. of TN/N | G-mean (%) | Sen (%) | Spec (%) |
|---|---|---|---|---|---|
| *Lesion level* | | | | | |
| TCFA | 17/19 | 53/69 | 82.9 | 89.5 | 76.8 |
| PB ≥ 70% | 17/18 | 52/70 | 83.8 | 94.4 | 74.3 |
| MLA ≤ 4 mm² | 21/22 | 56/66 | 90.0 | 95.5 | 84.8 |
| *Patient level* | | | | | |
| TCFA | 17/19 | 29/42 | 78.6 | 89.5 | 69.0 |
| PB ≥ 70% | 17/17 | 29/44 | 81.2 | 100.0 | 65.9 |
| MLA ≤ 4 mm² | 20/21 | 30/40 | 84.5 | 95.2 | 75.0 |

Notes: *TP*, true positive; *P*, positive samples; *TN*, true negative; *N*, negative samples.
Source: Reprinted with permission from Zhang L, Wahle A, Chen Z, Lopez JJ, Kovarnik T, Sonka M. Predicting locations of high-risk plaques in coronary arteries in patients receiving statin therapy. *IEEE Trans Med Imaging*. 2018;37(1):151–161. ©2018 IEEE.

plaque development and flagging locations of potential high-risk plaque. Any of the components (especially segmentation and registration) can be utilized independently for other plaque-related tasks and studies.

**GLOSSARY**

**IVUS,** Intravascular ultrasound—imaging modality using a probe inside a vessel

**VH,** Virtual histology—plaque characterization based on IVUS

**ThCFA,** Thick-cap fibro-atheroma—plaque phenotype with necrotic core enclosures

**TCFA,** Thin-cap fibro-atheroma—vulnerable plaque phenotype prone to rupture

**MACE,** Major adverse cardiac events—for example, myocardial infarction, heart failure

**LOGISMOS,** Layered optimal graph image segmentation for multiple objects and surfaces

**JEI,** Just enough interaction—LOGISMOS extension for efficient editing

**SVM,** Support vector machine—supervised linear classifier for prediction tasks

**LOPO,** Leave one patient out—cross-validation approach to maximize usage of available data

## ACKNOWLEDGMENTS

This study was supported in part by the grants from the National Institutes of Health, USA (R01HL063373, R01EB004640); Agency of Ministry of Health, Czech Republic (IGA NT13224-4/2012); and the National Natural Science Foundation of China (81501545, 61427806, 61372006).

## REFERENCES

1. Stone PH, Saito S, Takahashi S, et al. Prediction of progression of coronary artery disease and clinical outcomes using vascular profiling of endothelial shear stress and arterial plaque characteristics: the PREDICTION study. *Circulation.* 2012; 126(2):172–181.
2. Sanz J, Fayad ZA. Imaging of atherosclerotic cardiovascular disease. *Nature.* 2008; 451(7181):953–957.
3. Garcia-Garcia HM, Jang IK, Serruys PW, Kovacic JC, Narula J, Fayad ZA. Imaging plaques to predict and better manage patients with acute coronary events. *Circ Res.* 2014; 114(12):1904–1917.
4. Moore KJ, Tabas I. Macrophages in the pathogenesis of atherosclerosis. *Cell.* 2011; 145(3):341–355.
5. Nair A, Kuban BD, Tuzcu EM, Schoenhagen P, Nissen SE, Vince DG. Coronary plaque classification with intravascular ultrasound radiofrequency data analysis. *Circulation.* 2002; 106(17):2200–2206.
6. Zhang L, Wahle A, Chen Z, Lopez JJ, Kovarnik T, Sonka M. Predicting locations of high-risk plaques in coronary arteries in patients receiving statin therapy. *IEEE Trans Med Imaging.* 2018; 37(1):151–161.
7. Stary HC. Natural history and histological classification of atherosclerotic lesions. *Arterioscler Thromb Vasc Biol.* 2000; 20(5):1177–1178.
8. Kubo T, Maehara A, Mintz GS, et al. The dynamic nature of coronary artery lesion morphology assessed by serial virtual histology intravascular ultrasound tissue characterization. *J Am Coll Cardiol.* 2010; 55(15):1590–1597.
9. Zhao Z, Witzenbichler B, Mintz GS, et al. Dynamic nature of nonculprit coronary artery lesion morphology in STEMI: a serial IVUS analysis from the HORIZONS-AMI trial. *JACC Cardiovasc Imaging.* 2013; 6(1):86–95.
10. Chatzizisis YS, Jonas M, Coskun AU, et al. Prediction of the localization of high-risk coronary atherosclerotic plaques on the basis of low endothelial shear stress. *Circulation.* 2008; 117(8):993–1002.
11. Samady H, Eshtehardi P, McDaniel MC, et al. Coronary artery wall shear stress is associated with progression and transformation of atherosclerotic plaque and arterial remodeling in patients with coronary artery disease. *Circulation.* 2011; 124(7):779–788.
12. Corban MT, Eshtehardi P, Suo J, et al. Combination of plaque burden, wall shear stress, and plaque phenotype has incremental value for prediction of coronary atherosclerotic plaque progression and vulnerability. *Atherosclerosis.* 2014; 232(2):271–276.
13. Klingensmith JD, Schoenhagen P, Tajaddini A, et al. Automated three-dimensional assessment of coronary artery anatomy with intravascular ultrasound scanning. *Am Heart J.* 2003; 145(5):795–805.
14. Sonka M, Zhang X, Siebes M, et al. Segmentation of intravascular ultrasound images: a knowledge-based approach. *IEEE Trans Med Imaging.* 1995; 14(4):719–732.
15. Herrington DM, Johnson T, Santago P, Snyder WE. Semi-automated boundary detection for intravascular ultrasound. *Proc Computers in Cardiology 1992, Durham, NC.* Los Alamitos, CA: IEEE-CS Press; 1992:103–106.
16. Kass M, Witkin A, Terzopoulous D. Snakes: active contour models. *Int J Comput Vis.* 1988; 1(4):321–331.
17. Klingensmith JD, Shekhar R, Vince DG. Evaluation of three-dimensional segmentation algorithms for the identification of luminal and medial-adventitial borders in intravascular ultrasound images. *IEEE Trans Med Imaging.* 2000; 19(10):996–1011.
18. Wahle A, Lopez JJ, Olszewski ME, et al. Plaque development, vessel curvature, and wall shear stress in coronary arteries assessed by X-ray angiography and intravascular ultrasound. *Med Image Anal.* 2006; 10(4): 615–631.
19. Mojsilović A, Popović M, Amodaj M, Babić R, Ostojić M. Automatic segmentation of intravascular ultrasound images; a texture-based approach. *Ann Biomed Eng.* 1997; 25(6):1059–1071.
20. Zhang X, McKay CR, Sonka M. Tissue characterization in intravascular ultrasound images. *IEEE Trans Med Imaging.* 1998; 17(6):889–899.
21. Brusseau E, de Korte CL, Mastik F, Schaar J, van der Steen AFW. Fully automatic luminal contour segmentation in intracoronary ultrasound imaging—a statistical approach. *IEEE Trans Med Imaging.* 2004; 23(5): 554–566.
22. Li W, von Birgelen C, Di Mario C, et al. Semi-automatic contour detection for volumetric quantification of intravascular ultrasound. *Proc Computers in Cardiology 1994, Bethesda, MD.* Los Alamitos, CA: IEEE-CS Press; 1994/1995:277–280.
23. Sun S, Sonka M, Beichel RR. Graph-based IVUS segmentation with efficient computer-aided refinement. *IEEE Trans Med Imaging.* 2013; 32(8):1536–1549.
24. Timmins L, Suever J, Eshtehardi P, et al. Framework to co-register longitudinal virtual histology-intravascular ultrasound data in the circumferential direction. *IEEE Trans Med Imaging.* 2013; 32(11):1989–1996.
25. Zhang L, Downe RW, Chen Z, et al. Side-branch guided registration of intravascular ultrasound pullbacks in coronary arteries. In: *MICCAI Workshop in Computing and Visualization for Intravascular Imaging and Computer Assisted Stenting (CVII-STENT).* 2014:44–51.
26. Alberti M, Balocco S, Carrillo X, Mauri J, Radeva P. Automatic non-rigid temporal alignment of IVUS sequences. In: *Medical Image Computing and Computer-Assisted Intervention—MICCAI 2012.* Springer; 2012:642–650.

27. Alberti M, Balocco S, Carrillo X, Mauri J, Radeva P. Automatic non-rigid temporal alignment of intravascular ultrasound sequences: method and quantitative validation. *Ultrasound Med Biol.* 2013; 39(9):1698–1712.

28. Vukicevic AM, Stepanovic NM, Jovicic GR, Apostolovic SR, Filipovic ND. Computer methods for follow-up study of hemodynamic and disease progression in the stented coronary artery by fusing IVUS and X-ray angiography. *Med Biol Eng Comput.* 2014; 52(6):539–556.

29. Zhang L, Wahle A, Chen Z, et al. Simultaneous registration of location and orientation in intravascular ultrasound pullbacks pairs via 3D graph-based optimization. *IEEE Trans Med Imaging.* 2015; 34(12):2550–2561.

30. Zhou F, Torre F. Canonical time warping for alignment of human behavior. In: Bengio Y, Schuurmans D, Lafferty JD, Williams CKI, Culotta A, eds. *Advances in Neural Information Processing Systems.* Curran Associates, Inc. 2009:2286–2294.

31. Stone GW, Maehara A, Lansky AJ, et al. A prospective natural-history study of coronary atherosclerosis. *N Engl J Med.* 2011; 364(3):226–235.

32. Calvert PA, Obaid DR, O'Sullivan M, et al. Association between IVUS findings and adverse outcomes in patients with coronary artery disease. *JACC Cardiovasc Imaging.* 2011; 4(8):894–901.

33. Cheng JM, Garcia-Garcia HM, de Boer SPM, et al. In vivo detection of high-risk coronary plaques by radiofrequency intravascular ultrasound and cardiovascular outcome: results of the ATHEROREMO-IVUS study. *Eur Heart J.* 2014; 35(10):639–647.

34. Mintz GS, Nissen SE, Anderson WD, et al. American College of Cardiology clinical expert consensus document on standards for acquisition, measurement and reporting of intravascular ultrasound studies IVUS. *J Am Coll Cardiol.* 2001; 37(5):1478–1492.

35. Nair A, Margolis MP, Kuban BD, Vince DG. Automated coronary plaque characterisation with intravascular ultrasound backscatter: ex vivo validation. *EuroIntervention.* 2007; 3(1):113–120.

36. García-García HM, Mintz GS, Lerman A, et al. Tissue characterisation using intravascular radiofrequency data analysis: recommendations for acquisition, analysis, interpretation and reporting. *EuroIntervention.* 2009; 5(2): 177–189.

37. Nicholls SJ, Ballantyne CM, Barter PJ, et al. Effect of two intensive statin regimens on progression of coronary disease. *N Engl J Med.* 2011; 365(22):2078–2087.

38. Banach M, Serban C, Sahebkar A, et al. Impact of statin therapy on coronary plaque composition: a systematic review and meta-analysis of virtual histology intravascular ultrasound studies. *BMC Med.* 2015; 13(1):229.

39. Adamson PD, Dweck MR, Newby DE. The vulnerable atherosclerotic plaque: in vivo identification and potential therapeutic avenues. *Heart.* 2015; 101(21): 1755–1766.

40. Breiman L. Statistical modeling: the two cultures (with comments and a rejoinder by the author). *Stat Sci.* 2001; 16(3):199–231.

41. Gorodeski EZ, Ishwaran H, Kogalur UB, et al. Use of hundreds of electrocardiographic biomarkers for prediction of mortality in postmenopausal women. *Circ Cardiovasc Qual Outcomes.* 2011; 4(5):521–532.

42. Jordan MI, Mitchell TM. Machine learning: trends, perspectives, and prospects. *Science.* 2015; 349(6245): 255–260.

43. Wang Z, Jia H, Tian J, et al. Computer-aided image analysis algorithm to enhance in vivo diagnosis of plaque erosion by intravascular optical coherence tomography. *Circ Cardiovasc Imaging.* 2014; 7(5):805–810.

44. Zhang L, Wahle A, Chen Z, Lopez J, Kovarnik T, Sonka M. Prospective prediction of thin-cap fibroatheromas from baseline virtual histology intravascular ultrasound data. In: *International Conference on Medical Image Computing and Computer-Assisted Intervention.* Springer; 2015: 603–610.

45. Zhang L, Wahle A, Chen Z, Lopez J, Kovarnik T, Sonka M. Location-specific prediction of vulnerable plaque using IVUS, virtual histology, and spatial context. In: *IEEE 13th International Symposium on Biomedical Imaging (ISBI).* IEEE; 2016:1354–1358.

46. Kovarnik T, Mintz GS, Skalicka H, et al. Virtual histology evaluation of atherosclerosis regression during atorvastatin and ezetimibe administration. *Circ J.* 2012; 76(1):176–183.

47. Wu X, Chen DZ. Optimal net surface problems with applications. In: *Automata, Languages and Programming Lecture Notes in Computer Science;* vol. 2380:. Springer; 2002:775.

48. Li K, Wu X, Chen DZ, Sonka M. Globally optimal segmentation of interacting surfaces with geometric constraints. In: *Proceedings of the 2004 IEEE Computer Society Conference on Computer Vision and Pattern Recognition (CVPR).* 2004:394–399 vol. 1.

49. Li K, Wu X, Chen DZ, Sonka M. Optimal surface segmentation in volumetric images—a graph-theoretic approach. *IEEE Trans Pattern Anal Mach Intell.* 2006; 28(1): 119–134.

50. Yin Y, Zhang X, Williams R, Wu X, Anderson DD, Sonka M. LOGISMOS—layered optimal graph image segmentation of multiple objects and surfaces: cartilage segmentation in the knee joint. *IEEE Trans Med Imaging.* 2010; 29(12): 2023–2037.

51. Goldberg AV, Hed S, Kaplan H, Kohli P, Tarjan RE, Werneck RF. Faster and more dynamic maximum flow by incremental breadth-first search. In: *Algorithms-ESA 2015, Lecture Notes in Computer Science;* vol. 9294: Springer; 2015:619–630.

52. Zhang H, Kashyap S, Wahle A, Sonka M. Highly modular multi-platform development environment for automated segmentation and just enough interaction. In: *Interactive Medical Image Computing: 3rd International Workshop, IMIC 2016, Held in Conjunction With MICCAI 2016, Athens, Greece, October.* 2016.

53. Felzenszwalb PF, Zabih R. Dynamic programming and graph algorithms in computer vision. *IEEE Trans Pattern Anal Mach Intell.* 2011; 33(4):721–740.

54. Sonka M, Hlavac V, Boyle R. *Image Processing, Analysis, and Machine Vision*. Cengage Learning; 2014.

55. Wahle A, Prause GPM, DeJong SC, Sonka M. Geometrically correct 3-D reconstruction of intravascular ultrasound images by fusion with biplane angiography—methods and validation. *IEEE Trans Med Imaging*. 1999; 18(8): 686–699.

56. von Birgelen C, Hartmann M, Mintz GS, Baumgart D, Schmermund A, Erbel R. Relation between progression and regression of atherosclerotic left main coronary artery disease and serum cholesterol levels as assessed with serial long-term (≥12 months) follow-up intravascular ultrasound. *Circulation*. 2003; 108(22):2757–2762.

57. Glagov S, Weisenberg E, Zarins CK, Stankunavicius R, Kolettis GJ. Compensatory enlargement of human atherosclerotic coronary arteries. *N Engl J Med*. 1987; 316(22):1371–1375.

58. Pasterkamp G, Galis ZS, De Kleijn DPV. Expansive arterial remodeling: location, location, location. *Arterioscler Thromb Vasc Biol*. 2004; 24(4):650–657.

59. Hermiller JB, Tenaglia AN, Kisslo KB, et al. In vivo validation of compensatory enlargement of atherosclerotic coronary arteries. *Am J Cardiol*. 1993; 71(8):665–668.

60. Pasterkamp G, Wensing PJW, Post MJ, Hillen B, Mali WPTM, Borst C. Paradoxical arterial wall shrinkage may contribute to luminal narrowing of human atherosclerotic femoral arteries. *Circulation*. 1995; 91(5): 1444–1449.

61. Varnava AM, Mills PG, Davies MJ. Relationship between coronary artery remodeling and plaque vulnerability. *Circulation*. 2002; 105(8):939–943.

62. Rodriguez-Granillo GA, Serruys PW, García-García HM, et al. Coronary artery remodelling is related to plaque composition. *Heart*. 2006; 92(3):388–391.

63. Chen Z, Wahle A, Zhang L, Kovarnik T, Lopez JJ, Sonka M. Comprehensive serial study of dynamic remodeling of atherosclerotic coronary arteries using IVUS. In: *SPIE Medical Imaging 2016: Biomedical Applications in Molecular, Structural, and Functional Imaging*. 9788. SPIE; 2016 97880Y.1-97880Y.6.

64. Zarins CK, Weisenberg E, Kolettis G, Stankunavicius R, Glagov S. Differential enlargement of artery segments in response to enlarging atherosclerotic plaques. *J Vasc Surg*. 1988; 7(3):386–394.

65. Guyon I, Weston J, Barnhill S, Vapnik V. Gene selection for cancer classification using support vector machines. *Mach Learn*. 2002; 46(1-3):389–422.

66. Guyon I, Elisseeff A. An introduction to variable and feature selection. *J Mach Learn Res*. 2003; 3:1157–1182.

67. Cortes C, Vapnik V. Support-vector networks. *Mach Learn*. 1995; 20(3):273–297.

68. Chang CC, Lin CJ. LIBSVM: a library for support vector machines. *ACM Trans Intell Syst Technol (TIST)*. 2011; 2 (3):27.

69. Kovarnik T, Chen Z, Wahle A, et al. Pathologic intimal thickening plaque phenotype: not as innocent as previously thought. A serial 3D intravascular ultrasound virtual histology study. *Rev Esp Cardiol (English ed.)*. 2017; 70(1):25–33.

70. Virmani R, Burke AP, Farb A, Kolodgie FD. Pathology of the vulnerable plaque. *J Am Coll Cardiol*. 2006; 47(8 suppl): C13–C18.

71. Maehara A, Mintz GS, Bui AB, et al. Morphologic and angiographic features of coronary plaque rupture detected by intravascular ultrasound. *J Am Coll Cardiol*. 2002; 40(5):904–910.

72. Chong DY, Kim HJ, Lo P, et al. Robustness-driven feature selection in classification of fibrotic interstitial lung disease patterns in computed tomography using 3D texture features. *IEEE Trans Med Imaging*. 2016; 35 (1):144–157.

73. Torheim T, Malinen E, Kvaal K, et al. Classification of dynamic contrast enhanced MR images of cervical cancers using texture analysis and support vector machines. *IEEE Trans Med Imaging*. 2014; 33(8):1648–1656.

74. Haralick RM, Shanmugam K, Dinstein I. Textural features for image classification. *IEEE Trans Syst Man Cybern*. 1973; 3(6):610–621.

75. Akbani R, Kwek S, Japkowicz N. Applying support vector machines to imbalanced datasets. In: *European Conference on Machine Learning*. Springer; 2004:39–50.

76. He H, Garcia EA. Learning from imbalanced data. *IEEE Trans Knowl Data Eng*. 2009; 21(9):1263–1284.

# CHAPTER 9

# Training Convolutional Nets to Detect Calcified Plaque in IVUS Sequences

RICARDO ÑANCULEF[a] • PETIA RADEVA[b,c] • SIMONE BALOCCO[b,c]
[a]Federico Santa María Technical University, Valparaíso, Chile, [b]Department of Mathematics and Informatics, University of Barcelona, Barcelona, Spain, [c]Computer Vision Center, Bellaterra, Spain

## 1. INTRODUCTION

Intravascular ultrasound (IVUS) is a catheter-based imaging technique generally used during percutaneous interventions. In clinical practice, an IVUS acquisition consists in a set of frames obtained while the probe, inside the vessel, is pulled-back (pullback) at constant velocity (0.5–1.0 mm/s). As a result, a sequence of thousands of frames reproducing the internal vascular morphology is obtained (see Fig. 1). Despite the large amount of data in a pullback, physicians mainly focus on regions of the vessel characterized by clinically relevant observations ("clinical events") such as the presence of stenosis, characterized by calcifications or lipid pools, indicating potential threats to the patient health. An IVUS sequence can be considered as a connected series of clinical events (see Fig. 1). The quick identification of these regions is a fundamental task for the diagnosis of the patient clinical conditions and for guiding physicians in choosing the most suited treatment. The huge number of frames that need to be analyzed before finding the location and extension of such regions makes this procedure extremely time consuming.

Several researchers, inspired by the visual inspection performed by physicians, studied the automatic detection of clinical events in IVUS images using pattern recognition techniques. In 2011, a first approach for the automatic key frames detection in IVUS was proposed. The key frames are markers delimiting clinical events along the vessel. The detection was based on the analysis of vessel morphological profiles extracted using supervised classifiers.[1] The area measurements of the lumen, vessel, fibrotic, lipidic, and calcified plaque were extracted from the IVUS sequence. Then, the selection of key frames was obtained using the symbolic aggregate approximation (SAX) algorithm. In another study,[2] the automatic identification of vascular bifurcations was studied. The approach was based on a random forest classifier which computed textural descriptors along

angular sectors of the image. In Ref. 3, a method for automated selection of vessel segments containing culprit and nonculprit frames was presented. The method uses the LOESS nonparametric regression method to analyze the distribution of atherosclerotic plaque burden appearing in each IVUS frame. More recently, an approach for the detection of the boundaries and the position of the stent along the pullback has been proposed.[4] In that study, the measure of likelihood for a frame to contain a stent is computed using a stacked sequential learning classification approach, in which specific stent features were designed. Then, a robust binary representation of the presence of the stent in the pullback is obtained applying an iterative approximation of the signal using the SAX algorithm.

The main limitation of such approaches is that they require careful design of handcrafted features, which is specific for each clinical event. In general, these features describing the appearance of the target structures are costly effort to obtain and they are not sufficiently flexible to describe the complicated characteristics of some clinical events.

To tackle this restriction, we investigate a novel approach for IVUS sequence analysis based on convolutional neural nets (CNN). This technique is able to automatically learn the optimal set of features for the detection of a clinical event from a dataset where only one label per frame has been provided indicating presence of the pattern. In this chapter, the presence of arterial calcifications has been selected as an exemplar clinical event, since it plays a major role in the appearance and development of atherosclerosis.[1,3,5] Our contribution is hence a novel approach to automatically select the sections of a pullback which contain calcification and should be considered for a more detailed inspection. Our preliminary experiments using in vivo image sequences acquired from 80 patients show that convolutional architectures improve detections of a shallow classifier in terms of $F_1$-measure, precision, and recall.

*Intravascular Ultrasound.* https://doi.org/10.1016/B978-0-12-818833-0.00009-6

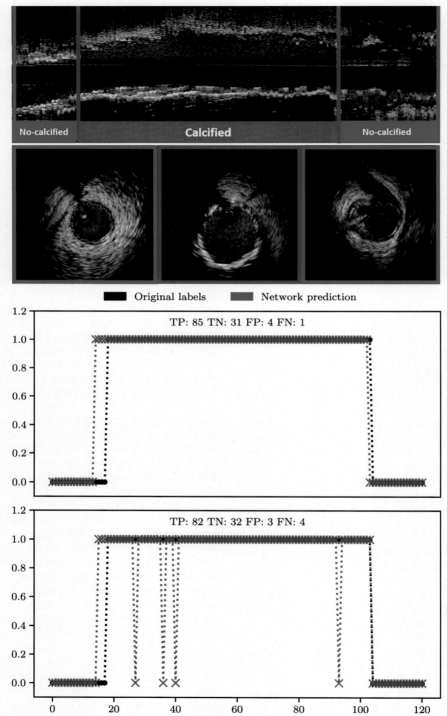

FIG. 1 Detection examples. *Top:* Longitudinal view of an exemplar IVUS pullback having a section containing calcifications. *Bottom:* Predictions obtained by the proposed pipeline, after step 2 (*bottom*) and after step 3 (*top*).

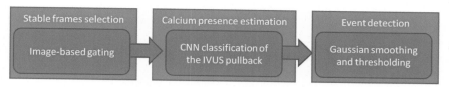

FIG. 2 Schematic representation of the frame selection technique.

## 2. METHODOLOGY

During an IVUS data acquisition, an ultrasonic probe is pulled-back inside the patient's vessel, producing a sequence of frames $(X_\ell) = (X_1, X_2, ..., X_L)$ called *a pullback*. Each IVUS frame is stored and processed as an $M \times N$ matrix $X_\ell$ that can reveal different conditions of the vessel, namely, the presence of fibrotic plaque, calcified plaque, lipid pool, implanted stent or bifurcation,[2,6,7] among others.

In this chapter, we address the problem of selecting the frames in a pullback $(X_\ell)$ that are relevant for the presence arterial calcification. Our framework has three components as depicted in Fig. 2. The first component implements a gating procedure that selects the most stable frames in the pullback, discarding images affected by artifacts due to the heart's contraction/expansion cycle.[8] The second component performs a frame-based analysis of the sequence, producing a temporal signal $(S_\ell) = (S_1, S_2, ..., S_L)$, where a high score $S_\ell \in [0, 1]$ indicates that relevant levels of calcification have been detected in $X_\ell$, while a low score indicates that the frame is not relevant for calcification analysis. The last component performs a sequence-based refinement of the temporal signal $(S_\ell)$, which aims to increase its temporal consistency. As detailed further, these corrected scores are used to select the final set of frames using an optimized thresholding procedure. An example of input and output of this detection system is provided in Fig. 2.

To implement the first step of this framework, we use the method introduced in Ref. 8. As for the other components, we adopt a supervised learning approach, that is, we consider a dataset $D = \{(X_\ell)_n\}$ of $N$ pullbacks that has been collected and annotated by a specialist. Ground-truth labels $T = \{(T_\ell)_n\}$ are assumed to be binary, that is, $T_{\ell, n} = 1$ if the frame $\ell$ of the $n$th pullback has been annotated for the presence of calcified plaque and 0 otherwise.

### 2.1. Frame-Based Detection

We implement the first step of our pipeline by training a convolutional net (CNN) to recognize calcified plaque on individual frames.

### 2.1.1. Convolutional architectures

Given an unknown function $f_0 : \mathbb{X} \to \mathbb{Y}$ that one needs to learn from data, neural networks implement a hypothesis $f : \mathbb{X} \to \mathbb{Y}$ that decomposes as the composition $f = f_1 \circ f_2 \circ \cdots f_M$ of more simple functions $f_m$ referred to as layers. In classic feed-forward nets (FFNs), layers receive as input a vector $a^{(m-1)}$ of size $I_{m-1}$ and compute as output a vector $a^{(m)}$ of size $I_m$, implementing a map of the form $a^{(m)} = g_m(W^{(m)} a^{(m-1)} + b^{(m)})$, where $W^{(m)}$ is a matrix of shape $I_m \times I_{m-1}$, $b \in \mathbb{R}^{I_m}$ and $g_m(\cdot)$ is a nonlinear function applied component-wise. Compared to FFNs, the early layers of a CNN allow two additional types of computation: convolution and pooling.

Convolutional layers receive as input an image $A^{(m-1)}$ (with $K_m$ channels) and compute as output a new image $A^{(m)}$ (composed of $O_m$ channels). The output at each channel is known as *a feature map*, and is computed as

$$A_o^{(m)} = g_m \left( \sum_k W_{ok}^{(m)} * A_k^{(m-1)} + b_o^{(m)} \right), \quad (1)$$

where $*$ denotes the (2D) convolution operation[a]

$$W_{ok} * A_k[s,t] = \sum_{p,q} A_k[s+p, t+q] W_{ok}[P-1-p, Q-1-q], \quad (2)$$

where $W_{ok}^{(m)}$ is matrix of shape $P_m \times Q_m$ and $b_o^{(m)} \in \mathbb{R}$. The matrix $W_{ok}^{(m)}$ parameterizes a spatial filter that the layer can use to detect or enhance some feature in the incoming image. The specific action of this filter is automatically learnt from data in the training process of the network.

Pooling layers of a CNN implement a spatial dimensionality reduction operation designed to reduce the number of trainable parameters for the next layers and allow them to focus on larger areas of the input pattern. Given an image $A^{(m-1)}$, a typical pooling layer with pool sizes $P_m, Q_m \in \mathbb{N}$, and strides $\alpha_m, \beta_m \in \mathbb{N}$ implements a channel-wise operation of the form

---

[a] The entries of a matrix $A$ are denoted as $A[i, j]$ to avoid an excessive number of subindices. Channels are still denoted using a subindex.

$$A_o^{(m)}[s,t] = \kappa \cdot \left( \sum_{p,q} (A_o^{(m-1)}[\alpha_m s + p, \beta_m t + q])^\rho \right)^{1/\rho}, \qquad (3)$$

where $\kappa, \rho \in \mathbb{N}$ are fixed parameters. Note that using $P_m = Q_m = \alpha_m = \beta_m$ corresponds to divide each channel of the input image into nonoverlapping $P_m \times Q_m$ patches and substitute the values in that region by a single value determined by $\rho$ and $\kappa$. In *max pooling* layers ($\rho = \infty$, $\kappa = 1$), this value is the maximum of the values found in the patch. In *average pooling* layers ($\rho = 1$, $\kappa = 1/PQ$), one takes the average of the values in the corresponding patch. The right choice of this function can make the model more robust to distortions in the input pattern.

Architecting a deep CNN stands for devising an appropriate succession of convolutional, pooling, and traditional (fully connected) layers, as well as their hyperparameters. As depicted in Fig. 3, typical architectures introduce a pooling layer after one or two convolutional layers, defining a convolutional block that is repeated until the size of feature map is small enough to introduce traditional layers. The transition from bidimensional (or multidimensional) layers to one-dimensional fully connected layers requires a special reshaping operation called "a flatten layer."

### 2.1.2. Training deep calcified plaque detectors

Given ground-truth labels for the presence of calcified plaque in IVUS images, we can train frame-wise detectors using a network $f(X)$, which predicts the probability

of an image $X$ containing the pattern. This is achieved by using a fully connected output layer with a sigmoid neuron, that is, a layer $f_M$ implementing the transformation $a^{(M)} = \sigma(W^{(M)} a^{(M-1)} + b^{(M)})$ with $\sigma(\xi) = (1 + \exp(\xi))^{-1}$. The network's parameters $\Theta$ can thus be learnt to maximize the log-likelihood

$$J(\Theta) = -\sum_n \sum_\ell Q\left( T_\ell^{(n)}, f_\Theta\left( X_\ell^{(n)} \right) \right), \qquad (4)$$

where $Q(y, p) = -(1 - y)(1 - p) - yp$ is known as the *cross-entropy* loss. The optimization of this objective function can be performed using back-propagation.[9] Essentially, the network parameters are iteratively adapted using a variant of stochastic gradient descent, where the derivatives corresponding to a matrix of unknown parameters at some layer $m$ are obtained from the derivatives of the next layer using the (multivariate) chain rule. In the case of the convolutional filters $W_{ok}^{(m)}$, this rule takes the form

$$\frac{\partial J}{\partial W_{ok}^{(m)}}[p,q] = \sum_{i,j} \frac{\partial J}{\partial A_o^{(m)}[i,j]} \frac{\partial A_o^{(m)}[i,j]}{\partial W_{ok}^{(m)}[p,q]},$$

$$\frac{\partial J}{\partial A_k^{(m-1)}}[s,t] = \sum_o \sum_{i,j} \frac{\partial J}{\partial A_o^{(m)}[i,j]} \frac{\partial A_o^{(m)}[i,j]}{\partial A_k^{(m-1)}[s,t]}. \qquad (5)$$

For deep networks, the success of this learning strategy heavily depends on the number and diversity of training instances. Since our dataset consists of few thousands of images, which is quite large for biomedical studies but

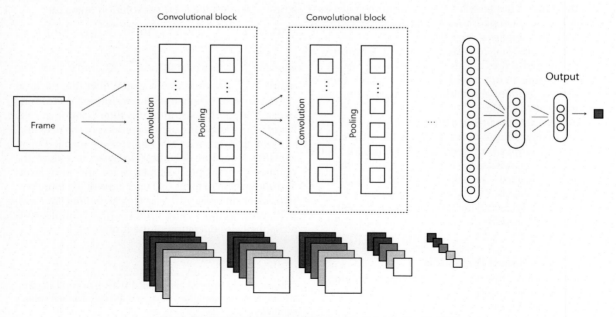

FIG. 3 Example of a CNN architecture popularized after AlexNet.

probably insufficient for training deep nets, we need to enrich the dataset with data augmentation strategies. Concretely, since we expect a high rotational invariance of the calcified plaque detectors, we expanded the original dataset by randomly rotating each image in the range 0–360 degrees. Each rotated image can also be flipped horizontally and vertically (examples are shown in Fig. 4). To make this process memory-efficient, we do not expand the dataset before training, but perform the data augmentation operations stochastically during back-propagation learning, that is, at each round of the training process (epoch), a random variant of the original dataset is processed by the network.

To make the training process even more robust to overfitting, we train the CNN using Dropout,[10] a regularization technique recently introduced to improve the generalization ability of deep nets. At each iteration of the learning process, an stochastic binary mask is applied to the output of each layer (each output feature map in the case of convolutional layers). In this way, the next layer is constrained to learn with incomplete information from the previous layer, preventing strong levels of coadaptation between the feature detectors extracted by the net. To properly handle the stochasticity of the learning signals arising from the use of Dropout and dynamic data augmentation, we used back-propagation with momentum and the bias correction method proposed in Ref. 11.

## 2.2. Improving Temporal Consistency

Classifying each frame in a pullback independently may be suboptimal for accurate detection because it does not properly enforce consistency of predicted labels over time. Ground-truth labels in contrast are temporally consistent, in the sense that the detection of the clinical pattern at time $t$ significantly increases the probability of observing the pattern in the next and the previous frames. Incorporation of such prior knowledge into the system requires a sequence-based analysis that can be carried out in many different ways, for example, applying a low-pass filter, introducing a regularizer or penalty into the learner's training criterion, or equipping the network with a form of "memory" about its previous states. In this chapter, we explore the simplest and more efficient of these alternatives, namely, we convolve the probabilities resulting from the first step of our pipeline with a Gaussian filter of width $\sigma$. Formally, if we denote by $(S_\ell)$ the sequence of predictions obtained by applying the CNN to the frames of a given pullback $(X_\ell)$, the second step of our pipeline is the sequence $(S_\ell')$ obtained as

$$S_\ell' = \sum_{t=0}^{T-1} S_{\ell+t} K_{T-1-t},$$ (6)

where $(K_\ell)$ is the sequence obtained by considering a discrete length-$T$ approximation of the Gaussian kernel. To obtain a sequence $(S_\ell')$ of the same length as the input sequence, $(S_\ell)$ is first zero-padded conveniently.

## 2.3. Final Selection of Relevant Frames

The network used to implement the first step of our pipeline has been trained to minimize the cross-entropy loss, that is, to approximate the posterior distribution

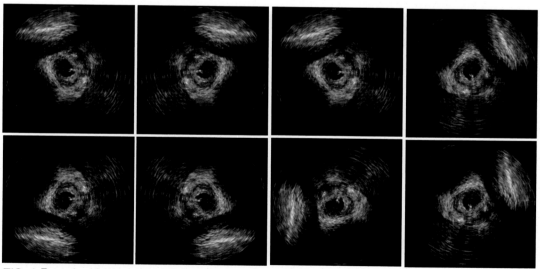

FIG. 4 Examples of data augmentation used to expand the original dataset. On *left* is the four possible flips and on the *right* four random rotations.

over the set of labels (0/1). In the context of binary classification, these probabilities typically translate into a decision by using the rule $Y(X) = I(f(X) \leq 0.5) = I(f(X) \leq 1 - f(X))$, which is known to be optimal when $f(X)$ is the true posterior and the classification accuracy is the metric to be optimized. However, in the context of pattern detection on biomedical data, two most frequent measures of performance are precision and recall.

Precision can be defined as the probability that the pattern of interest is present in the instances in which the system has detected the pattern. Recall corresponds to the probability that system detects the pattern of interest in the instances where this is indeed present. More formally, if $T(X)$ denotes the correct label of an instance $X$, and $S$ denotes a collection of instances, the following sets can be defined:

- True positives: $TP = |\{X \in S : Y(X) = 1 \text{ and } T(X) = 1\}|$.
- False positives: $FP = |\{X \in S : Y(X) = 1 \text{ and } T(X) = 0\}|$.
- False negatives: $TN = |\{X \in S : Y(X) = 0 \text{ and } T(X) = 1\}|$.
- True negatives: $TN = |\{X \in S : Y(X) = 0 \text{ and } T(X) = 0\}|$.

Precision and recall are thus computed as $P = TP/(TP + FP)$ and $R = TP/(TP + FN)$, respectively. The two quantities clearly trade-off against one another: it is always possible to get a perfect recall $R = 1$ at expenses of a very low precision and vice versa. A widely used metric that combines both scores is the $F_1$-measure defined as harmonic mean between precision and recall $F_1 = 2PR/(P + R)$. Compared to accuracy, $F_1$ increases only marginally as a function of TN giving much more relevance to TP. As shown in Ref. 12, for constant FP, $F_1$ is concave in TP while accuracy is just linear (lower returns). For fixed FN, in contrast, $F_1$ is convex in TN while accuracy is linear (larger returns). For this reason, optimal decision rules for accuracy can be far from optimal for $F_1$. For instance, Ref. 12 has shown that if one considers parameterized decision rules of the form $Y(X) = I(f(X) \leq \theta)$, the optimal threshold $\theta$ for $F_1$ is half the optimal $F_1$-score. This shows that the usual threshold $\theta = 0.5$ is optimal only if the rule $Y(X) = I(f(X) \leq 0.5)$ achieves a perfect $F_1$ of 1.

Unfortunately, training $f(X)$ to directly maximize $F_1$ is often much more difficult than optimizing the cross-entropy loss. We use here a more simple heuristic which consists in choosing $\theta$ to maximize the empirical $F_1$-measure on the training set, after the posterior probabilities $f(X)$ has been estimated. Though simple, this method has been found to be effective and is commonly applied in many scenarios such as document retrieval.[13] In addition, recent research has shown that a maximizer of the $F_1$-measure with the form $Y(X) = I(f(X) \leq \theta)$ always exist and its estimation is statistically consistent.[14] This means that the heuristic converges to the optimal rule as the size of the training set increases.

## 3. EXPERIMENTS AND DISCUSSION
### 3.1. Materials
A set of IVUS pullbacks corresponding to 80 patients was acquired from the "German Trias i Pujol" Hospital in Badalona (Spain) using an iLab echograph with a 40-MHz catheter and a pullback speed of 0.5 mm/s. The data acquisition protocol was approved by the IRB of each clinical center. The obtained sequences were variable in length, with an average of 111.8 frames. A total of 8914 IVUS images were obtained. One expert manually annotated each image for the presence of calcification. The fraction of frames containing the pattern (positives) was 40.65%.

To study the performance of the methods described in this chapter, we randomly split the dataset into three disjoint subsets: a training, a validation, and a test subset. The model is built with data from the training subset only. Data in the validation set is used to choose among models and alternative configurations. The test set is used to evaluate the performance of the model on novel/unseen cases. As illustrated in Fig. 5, there are two different ways to compute the partition. In the first method, images are randomly drawn from the entire set of images. With high probability, this method distributes images of all the patients into the different subsets. The second method, in contrast, draws entire image sequences into a subset. In this method, the images of a patient can belong to only one of the subsets in the partition. Therefore, images used for testing and images used for training differ more significantly.

For the experiments of this section, we have adopted the second approach since, in practice, the system should be able to handle images from patients that have not been observed previously. In addition, to allow a statistical significance analysis of the results, we evaluate the methods using 10 different partitions of the dataset. For each split, we reserved 36 pullbacks for training (4024 images on average), 8 pullbacks for validation (1006 images on average), and 36 pullbacks for testing (4024 images on average).

### 3.2. Evaluation Metrics
Methods will be evaluated in terms of precision, recall, and F-score. These metrics have been defined in Section 2.3 considering a generic collection of frames. However, as our dataset is a collection of *variable-length frame sequences* $D = \{(X_\ell^n)\}$, there are two different ways to compute these metrics in practice. As illustrated in Fig. 6, these methods lead to different results and different interpretations of the results.

1. **Pullback-based statistics.** This method computes precision, recall, and F-score for a given sequence

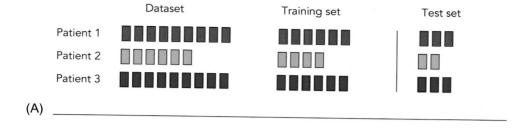

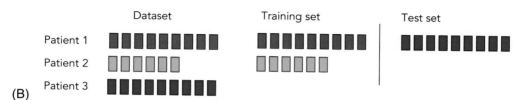

FIG. 5 Train/test splits. The figure illustrates two different ways to split the dataset into a training subset and a test subset. In (A), the training set is obtained by sampling the entire set of images, and thus it includes example frames from all the patients. In (B), the training set is obtained by sampling entire pullbacks from the dataset and thus a patient can only belong to one of the two subsets. The second setting is more challenging, because the model has never seen images from patients in the test set.

(patient) and then averages among the sequences of the dataset. In this method, the total number of positives, the total number of predicted positives, and, in general, the system's confusion matrix changes from sequence to sequence. To state it more clearly, if $Y(X_\ell^n)$ denotes the prediction on frame $\ell$ of sequence $n$, and $T_\ell^n$ is the ground-truth label, then this method computes the precision, recall, and F-score corresponding to the $n$th sequence in the dataset as

$$P_n = \frac{\text{TP}_n}{\text{TP}_n + \text{FP}_n}, \quad R_n = \frac{\text{TP}_n}{\text{TP}_n + \text{FN}_n|}, \quad F_n = \frac{2 P_n R_n}{P_n + R_n}, \quad (7)$$

respectively, where

$$\begin{aligned} \text{TP}_n &= \sum_\ell I(Y(X_\ell^n) = 1 \text{ and } T_\ell^n = 1), \\ \text{FP}_n &= \sum_\ell I(Y(X_\ell^n) = 1 \text{ and } T_\ell^n = 0), \\ \text{TN}_n &= \sum_\ell I(Y(X_\ell^n) = 0 \text{ and } T_\ell^n = 0), \\ \text{FN}_n &= \sum_\ell I(Y(X_\ell^n) = 0 \text{ and } T_\ell^n = 1). \end{aligned} \quad (8)$$

Dataset scores are obtained as

$$P = \frac{1}{N} \sum_n P_n, \quad R = \frac{1}{N} \sum_n R_n, \quad F_1 = \frac{1}{N} \sum_n F_n, \quad (9)$$

where $N$ is the total number of sequences.

2. **Frame-based statistics.** This approach computes precision, recall, and F-score considering all the frames in the dataset as a whole, ignoring the fact that they belong to different sequences (patients). This can be achieved by counting the number of true positives, false positives, true negatives, and false negatives obtained among the entire dataset:

$$\begin{aligned} \text{TP} &= \sum_n \sum_\ell I(Y(X_\ell^n) = 1 \text{ and } T_\ell^n = 1), \\ \text{FP} &= \sum_n \sum_\ell I(Y(X_\ell^n) = 1 \text{ and } T_\ell^n = 0), \\ \text{TN} &= \sum_n \sum_\ell I(Y(X_\ell^n) = 0 \text{ and } T_\ell^n = 0), \\ \text{FN} &= \sum_n \sum_\ell I(Y(X_\ell^n) = 0 \text{ and } T_\ell^n = 1). \end{aligned} \quad (10)$$

Dataset scores are obtained by applying the usual formulae for precision, recall, and F-score to these global counters

$$P = \frac{\text{TP}}{\text{TP} + \text{FP}}, \quad R = \frac{\text{TP}}{\text{TP} + \text{FN}}, \quad F = \frac{2PR}{P + R}. \quad (11)$$

Note that a precision score obtained using the pullback-based approach can be interpreted as the expected proportion of calcium-detected frames in a sequence that are actual positives. In contrast, a precision score computed using the frame-based approach

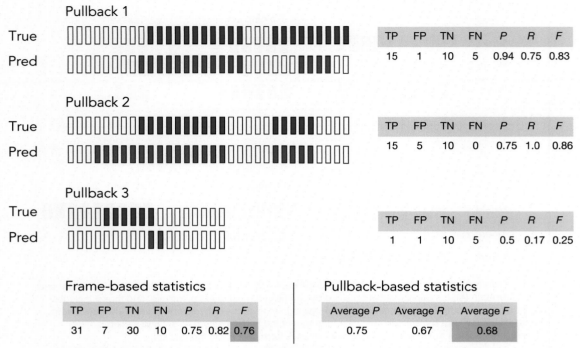

FIG. 6 Pullback-based versus frame-based statistics. The figure illustrates two different ways to compute precision, recall, and $F$-score in a dataset of variable-length sequences. Method (A) leads to the frame-based statistics defined by Eqs. (10), (11). Method (B) leads to the pullback-based statistics defined by Eqs. (7), (9) and discussed in the text.

should be interpreted as the probability that a *single* calcium-detected frame is an actual positive. If the number of positives and negatives varies significantly from sequence to sequence, these two quantities can be quite different. Similarly, a recall score obtained using the first method can be interpreted as the expected proportion of calcified frames in a pullback that are actually recognized as positives by the system. The recall score computed by the second method instead should be interpreted as the probability that a *single* calcified frame is correctly detected as a positive. In general, pullback-based scores are estimates of the performance of the system on the entire *sequences* of frames. Frame-based scores are instead estimates of the performance of the system on *single* frames. For this reason, the results we reported here are always pullback-based scores.

A final remark regarding the metrics reported in this work has to do with limit cases in which the denominator for $R_n$ in Eq. (7) is zero or the denominator for $P_n$ in Eq. (7) is zero. We handle these cases in the following ways:

- $TP_n + FN_n = 0$ and $TP_n + FP_n \neq 0$. In this case, we have $TP_n = 0$ and $FP_n \neq 0$, which means that $P_n = TP_n/(TP_n + FP_n) = 0$. In this case, we set $R_n = 1$, but since $P_n = 0$, the $F$-score is $F_n = 0$.
- $TP_n + FN_n \neq 0$ and $TP_n + FP_n = 0$. In this case, we have $TP_n = 0$ and $FN_n \neq 0$, which means that $R_n = TP_n/(TP_n + FN_n) = 0$. In this case, we set $P_n = 1$, but since $R_n = 0$, the $F$-score is $F_n = 0$.
- $TP_n + FN_n = 0$ and $TP_n + FP_n = 0$. In this case, we have $TP_n = 0$, $FN_n = 0$, and $FP_n \neq 0$. Assuming that the sequence is not empty, this means that the sequence contains healthy frames that have been correctly recognized. Since this is an optimal prediction, we set $P_n = 1$, $R_n = 1$, and $F_n = 1$.

### 3.3. Experiments

Our main goal in this section is to assess (i) the efficacy of convolutional features for calcified plaque detection in IVUS frame sequences and (ii) the relevance of performing a sequence-based refinement of the CNN predictions. In this section, we evaluate different CNN configurations by varying the number of convolutional blocks and filters in each layer. The basic convolutional block consists of a convolutional layer with rectifier linear units followed by a standard max-pooling layer with pooling size $2 \times 2$ and stride $2 \times 2$. Our preliminary experiments revealed that small $3 \times 3$ filters for the convolutional layers produced always results better than or similar to larger filters, except for the shallowest architecture of only one convolutional block. As smaller filters reduce the number of parameters in the network, we fix the filter size to $3 \times 3$ for nets of depth $M > 1$ and determine the number of filters using the following protocol. For a net of $M$ convolutional blocks and $S$ filters in the first layer, the number of filters was reduced to the half every $\lfloor M/2 \rfloor$ layers. Importantly, architectural decisions were always taken observing the performance on the training and validation sets.

We trained the neural nets using the adaptive stochastic gradient descent method proposed in Ref. 11. The objective function is, as usual, the average crossentropy among the training examples. Following author's recommendations, we used a base step size of $\alpha = 10^{-3}$ and exponential decay rates of $\beta_1 = 0.9$ and $\beta_2 = 0.999$ for the first and second moment estimates of the gradient, respectively. For simplicity, we used a fixed number of 100 SGD iterations (epochs), thought the validation error would increase only by a small amount after epoch 20. Parameters $\sigma$ and $\theta$ used in steps 2 and 3 of our pipeline were determined using a simple $100 \times 100$ grid search with a range of $[0.1, 8]$ for $\sigma$ and $[0.1, 0.9]$ for $\theta$. The combination of values producing the best performance in the validation set were

finally selected. Statistics on the test set were observed only after the nets were selected and trained. Similarly, for the baseline, a linear SVM trained directly on the pixels of the IVUS images, the regularization parameter $C$ was selected using a logarithmic grid search on the interval $[10^{-5}, 10^3]$ using the validation set to select the best values.

Given a set of pullbacks (training, validation, or test set), calcified plaque detections are evaluated using $F_1$-measure, defined in Section 2, as the main performance metric. There are, however, two different ways to calculate this score in practice. The first approach consists in computing precision, recall, and $F_1$ separately for each sequence of frames and then aggregating the results among the different sequences. The second approach consists in computing the metrics among all the frames in the collection, as if they belonged to a single big sequence. This is similar to the difference between macroaverage and microaverage in multilabel classification. Obtaining high scores with the first method is, in general, much more challenging when the number of positives varies among sequences. In the following section, these approaches are referred to as *pullback-based* and *frame-based*, respectively.

## 3.4. Results

Table 1 summarizes the (average) $F_1$-score achieved by the different architectures considered for implementing the CNN-based detectors. The best result ($F_1 = 0.66$) was obtained using a CNN of five convolutional blocks, the deepest of the models considered in our

**TABLE 1**
$F_1$-Score of the CNN and the Full Pipeline.

| Model | TRAIN | | | | TEST | | | |
| | CNN | Pipeline | %Imp. | T | CNN | Pipeline | %Imp. | T |
|---|---|---|---|---|---|---|---|---|
| CNN1[a] | 0.86 (0.02) | 0.93 (0.01) | +0.08 | ≠ | 0.48 (0.04) | 0.56 (0.05) | +0.17 | ≠ |
| CNN1[b] | 0.86 (0.04) | 0.94 (0.01) | +0.09 | ≠ | 0.46 (0.05) | 0.53 (0.06) | +0.15 | ≠ |
| CNN2[a] | 0.84 (0.02) | 0.91 (0.01) | +0.08 | ≠ | 0.56 (0.04) | 0.60 (0.05) | +0.07 | ≠ |
| CNN3[a] | 0.87 (0.03) | 0.92 (0.02) | +0.05 | ≠ | 0.57 (0.04) | 0.61 (0.03) | +0.07 | ≠ |
| CNN3[b] | 0.88 (0.02) | 0.92 (0.02) | +0.05 | ≠ | 0.56 (0.03) | 0.62 (0.03) | +0.10 | ≠ |
| CNN4[a] | 0.87 (0.03) | 0.92 (0.01) | +0.06 | ≠ | 0.58 (0.07) | 0.63 (0.04) | +0.08 | ≠ |
| CNN4[b] | 0.90 (0.05) | 0.96 (0.01) | +0.06 | ≠ | 0.53 (0.05) | 0.58 (0.04) | +0.09 | ≠ |
| CNN4[c] | 0.69 (0.14) | 0.89 (0.02) | +0.29 | ≠ | 0.48 (0.12) | 0.63 (0.05) | +0.32 | ≠ |
| CNN5[a] | 0.71 (0.13) | 0.88 (0.02) | +0.23 | ≠ | 0.53 (0.08) | 0.66 (0.04) | +0.23 | ≠ |

*Notes*: Average $F_1$-score obtained by the full pipeline compared to the score obtained by the CNN. The column labeled as %Imp. shows the relative improvement in the pipeline with respect to the results of the CNN only. The column labeled as T indicates the result of a paired *t*-test computed with a significance level of $\alpha = 0.1$. Averages were computed among the 10 train/test splits. Standard deviations are in parenthesis. For a single train/test split, the median among the sequences was computed to wave the effect of outliers.

**TABLE 2**
**Precision of the CNN and the Full Pipeline.**

| | PRECISION | | | | | | | |
| | TRAIN | | | | TEST | | | |
| Model | CNN | Pipeline | %Imp. | T | CNN | Pipeline | %Imp. | T |
|---|---|---|---|---|---|---|---|---|
| CNN1[a] | 0.92 (0.03) | 0.92 (0.02) | +0.01 | = | 0.48 (0.03) | 0.63 (0.07) | +0.11 | ≠ |
| CNN1[b] | 0.93 (0.05) | 0.96 (0.03) | +0.03 | = | 0.46 (0.06) | 0.64 (0.05) | +0.06 | = |
| CNN2[a] | 0.91 (0.03) | 0.90 (0.04) | −0.00 | = | 0.56 (0.05) | 0.72 (0.07) | +0.07 | ≠ |
| CNN3[a] | 0.92 (0.02) | 0.93 (0.03) | +0.01 | = | 0.57 (0.05) | 0.74 (0.09) | +0.07 | ≠ |
| CNN3[b] | 0.94 (0.02) | 0.93 (0.02) | −0.00 | = | 0.56 (0.05) | 0.74 (0.04) | +0.06 | ≠ |
| CNN4[a] | 0.91 (0.04) | 0.93 (0.03) | +0.02 | = | 0.58 (0.06) | 0.75 (0.04) | +0.06 | ≠ |
| CNN4[b] | 0.95 (0.04) | 0.97 (0.02) | +0.02 | = | 0.53 (0.09) | 0.73 (0.09) | +0.06 | = |
| CNN4[c] | 0.97 (0.04) | 0.89 (0.04) | −0.09 | ≠ | 0.48 (0.08) | 0.76 (0.06) | −0.15 | ≠ |
| CNN5[a] | 0.97 (0.04) | 0.86 (0.05) | −0.12 | ≠ | 0.53 (0.07) | 0.77 (0.07) | −0.12 | ≠ |

*Notes:* Average precision obtained by the entire pipeline compared to the precision obtained by the CNN. Results were computed and formatted as in Table 1.

experiments. As presented in Table 2, this model achieves an average precision of 77% and an average recall of 81%, which correspond to the best precision and best recall observed among the architectures we tested. It is worth noting that the harmonic mean corresponding to a precision of 0.77 and a recall of 0.81 is 0.79, quite above the $F_1$ of 0.66 reported in the table. This confirms that assessing the detector on each pullback/patient separately, for averaging the results, is more challenging than assessing the detector using all the frames/images as a whole. The first setting (used in our experiments) favors detectors that achieve a good balance between precision and recall in *each* pullback.

As summarized in Table 3, a paired *t*-test revealed statistically significant differences between the $F_1$-measure of our best model (CNN5[a], five convolutional layers) and all the others, except models CNN4[a] and CNN4[c], both with four convolutional layers. This result may suggest that the deeper the model, the better the performance. The other results in Table 3 suggest, however, that depth is neither a necessary nor a sufficient condition for obtaining good performance. For instance, no significant differences were found between the $F_1$-score of model CNN2[a] of two convolutional layers and that of models CNN4[a] and CNN4[c]. On the other hand, model CNN4[b] (four convolutional layer) achieves an $F_1$ of 0.58, well below the average and is not significantly different from the score obtained by

the model CNN1[a] (one convolutional layer). These results could be explained, at least in part, by considering the small number of images that were available to train the models (36 patients, 4024 images). Even if (see Table 4) an increase in depth does not necessarily translate into an increase in the number of parameters, it is often argued that the expressive power of a deep model increases exponentially with the number of layers, making it more prone to overfitting.

One of the techniques that had a major impact on the performance of the pipeline was data augmentation (DA), since it allowed to deal with the issue of having a limited number of labeled images. Indeed, in Table 5, we compare the performance of all the models with and without using DA. It can be noted that this technique improves the $F_1$-score of all the architectures in the test set (9/9 statistically significant) with an average increase of 19%. As expected, the effect tends to be more relevant for the deepest models. CNN1[a] and CNN1[b] (one convolutional layer) improved their test score by only 4%, while CNN5[a] by a 19%. The effect of this technique on the training performance is exactly the opposite: DA reduces the $F_1$-score of all the architectures in the train set (9/9 statistically significant). Together, all these results demonstrate that DA acts as a regularizer, reducing overfitting, but it is exploited better by the deepest architectures, not necessarily those with more parameters. Tables 6 and 7 show that DA has a more

**TABLE 3**
**One-vs-One Statistical Analysis.**

| Model | $1^a$ | $1^b$ | $2^a$ | $3^a$ | $3^b$ | $4^a$ | $4^b$ | $4^c$ | $5^a$ |
|---|---|---|---|---|---|---|---|---|---|
| **CNN ONLY** | | | | | | | | | |
| CNN1$^a$ | = | = | ≠ | ≠ | ≠ | ≠ | ≠ | = | ≠ |
| CNN1$^b$ | = | = | ≠ | ≠ | ≠ | ≠ | ≠ | = | ≠ |
| CNN2$^a$ | ≠ | ≠ | = | = | = | = | = | ≠ | = |
| CNN3$^a$ | ≠ | ≠ | = | = | = | = | = | ≠ | = |
| CNN3$^b$ | ≠ | ≠ | = | = | = | = | = | ≠ | = |
| CNN4$^a$ | ≠ | ≠ | = | = | = | = | ≠ | ≠ | = |
| CNN4$^b$ | ≠ | ≠ | = | = | = | ≠ | = | ≠ | = |
| CNN4$^c$ | = | = | ≠ | ≠ | ≠ | ≠ | ≠ | = | = |
| CNN5$^a$ | ≠ | ≠ | = | = | = | = | = | = | = |
| **FULL PIPELINE** | | | | | | | | | |
| CNN1$^a$ | = | = | ≠ | ≠ | ≠ | ≠ | = | ≠ | ≠ |
| CNN1$^b$ | = | = | ≠ | ≠ | ≠ | ≠ | ≠ | ≠ | ≠ |
| CNN2$^a$ | ≠ | ≠ | = | = | = | ≠ | = | = | = |
| CNN3$^a$ | ≠ | ≠ | = | = | = | = | ≠ | = | ≠ |
| CNN3$^b$ | ≠ | ≠ | = | = | = | = | ≠ | = | ≠ |
| CNN4$^a$ | ≠ | ≠ | ≠ | = | = | = | ≠ | = | = |
| CNN4$^b$ | = | ≠ | = | ≠ | ≠ | ≠ | = | ≠ | ≠ |
| CNN4$^c$ | ≠ | ≠ | = | = | = | = | ≠ | = | = |
| CNN5$^a$ | ≠ | ≠ | ≠ | ≠ | ≠ | = | ≠ | = | = |

Notes: Conclusions of paired $t$-tests comparing the $F_1$-score (on the test set) achieved by the different architectures. A significance level of $\alpha = 0.1$ was applied. The null hypothesis was that the pair or results are equal against the alternative that the results are different.

**TABLE 4**
**Architectures Explored to Build the Detectors.**

| Model | No. of blocks | No. of filters | No. of parameters | Dropout |
|---|---|---|---|---|
| CNN1$^a$ | 1 | 16 | 63.665 | 0.1 |
| CNN1$^b$ | 1 | 32 | 127.329 | 0.1 |
| CNN2$^a$ | 2 | 64 + 32 | 47.905 | 0.1 |
| CNN3$^a$ | 3 | 64 + 32 + 32 | 34.625 | 0.1 |
| CNN3$^b$ | 3 | 64 + 64 + 32 | 62.305 | 0.1 |
| CNN4$^a$ | 4 | 32 + 32 + 16 + 16 | 17.089 | 0.1 |
| CNN4$^b$ | 4 | 64 + 64 + 32 + 32 | 66.433 | 0.1 |
| CNN4$^c$ | 4 | 64 + 64 + 32 + 32 | 66.433 | 0.4 |
| CNN5$^a$ | 5 | 64+64+32+32+16 | 70.385 | 0.4 |

Notes: Number of convolutional blocks, convolutional filters, and total number of trainable parameters in the CNN models evaluated in this section.

**TABLE 5**
**Effect of Data Augmentation on the $F_1$-Score.**

| | TRAIN | | | | TEST | | | |
|---|---|---|---|---|---|---|---|---|
| Model | WDA | DA | %Imp. | T | CNN | Pipeline | %Imp. | T |
| CNN1[a] | 1.00 (0.00) | 0.86 (0.02) | −0.14 | ≠ | 0.46 (0.03) | 0.48 (0.04) | +0.04 | = |
| CNN1[b] | 1.00 (0.00) | 0.86 (0.04) | −0.14 | ≠ | 0.44 (0.03) | 0.46 (0.05) | +0.04 | = |
| CNN2[a] | 1.00 (0.00) | 0.84 (0.02) | −0.16 | ≠ | 0.51 (0.02) | 0.56 (0.04) | +0.09 | ≠ |
| CNN3[a] | 1.00 (0.00) | 0.87 (0.03) | −0.13 | ≠ | 0.49 (0.03) | 0.57 (0.04) | +0.17 | ≠ |
| CNN3[b] | 1.00 (0.00) | 0.88 (0.02) | −0.12 | ≠ | 0.51 (0.04) | 0.56 (0.03) | +0.10 | ≠ |
| CNN4[a] | 0.99 (0.01) | 0.87 (0.03) | −0.13 | ≠ | 0.49 (0.07) | 0.58 (0.07) | +0.20 | ≠ |
| CNN4[b] | 1.00 (0.00) | 0.90 (0.05) | −0.10 | ≠ | 0.49 (0.05) | 0.53 (0.05) | +0.09 | ≠ |
| CNN4[c] | 0.89 (0.07) | 0.69 (0.14) | −0.23 | ≠ | 0.41 (0.11) | 0.48 (0.12) | +0.16 | = |
| CNN5[a] | 0.85 (0.14) | 0.71 (0.13) | −0.16 | ≠ | 0.45 (0.11) | 0.53 (0.08) | +0.19 | ≠ |

Notes: Comparison of the average $F_1$-score obtained by the pipeline trained with the data augmentation strategy (column DA) and without it (column WDA). The column labeled as %Imp. shows the relative improvement due to this strategy. The column labeled as T indicates the result of a paired $t$-test computed with a significance level of $\alpha = 0.1$. Averages were computed among the 10 train/test splits. Standard deviations are in parenthesis. For a single train/test split, the median among the sequences was computed to wave the effect of outliers.

**TABLE 6**
**Effect of Data Augmentation on Precision.**

| | TRAIN | | | | TEST | | | |
|---|---|---|---|---|---|---|---|---|
| Model | WDA | DA | %Imp. | T | CNN | Pipeline | %Imp. | T |
| CNN1[a] | 1.00 (0.00) | 0.92 (0.03) | −0.08 | ≠ | 0.46 (0.02) | 0.57 (0.03) | +0.22 | ≠ |
| CNN1[b] | 1.00 (0.00) | 0.93 (0.05) | −0.07 | ≠ | 0.41 (0.04) | 0.60 (0.06) | +0.46 | ≠ |
| CNN2[a] | 1.00 (0.00) | 0.91 (0.03) | −0.09 | ≠ | 0.53 (0.03) | 0.67 (0.05) | +0.27 | ≠ |
| CNN3[a] | 1.00 (0.00) | 0.92 (0.02) | −0.08 | ≠ | 0.54 (0.03) | 0.69 (0.05) | +0.28 | ≠ |
| CNN3[b] | 1.00 (0.00) | 0.94 (0.02) | −0.06 | ≠ | 0.55 (0.04) | 0.70 (0.05) | +0.27 | ≠ |
| CNN4[a] | 0.99 (0.01) | 0.91 (0.04) | −0.08 | ≠ | 0.61 (0.05) | 0.70 (0.06) | +0.16 | ≠ |
| CNN4[b] | 1.00 (0.00) | 0.95 (0.04) | −0.05 | ≠ | 0.57 (0.02) | 0.69 (0.09) | +0.21 | ≠ |
| CNN4[c] | 1.00 (0.00) | 0.97 (0.04) | −0.03 | ≠ | 0.77 (0.12) | 0.90 (0.08) | +0.17 | ≠ |
| CNN5[a] | 1.00 (0.00) | 0.97 (0.04) | −0.03 | ≠ | 0.85 (0.11) | 0.88 (0.07) | +0.04 | = |

Notes: Comparison of the (average) precision obtained by the pipeline trained with the data augmentation strategy (column DA) and without it (column WDA). Results were computed and formatted as in Table 5.

significant effect on precision than on recall. Precision improved by 39% with respect to the performance of the detectors trained without DA, but recall improved by only a 5% (on average among the models).

From Tables 1–8, we can conclude that the components of the pipeline devised to improve the temporal consistency of the detections provided by the CNN has a more significant effect on recall than on precision. Recall in the test set improved in all the cases (7/9

statistically significant) with an average increase of 19%. In contrast, precision in the test set, sometimes decreases, producing an average loss of 1.5%. For the model providing the best $F_1$-score (CNN5[a]), the last steps of the pipeline improve recall by 66% at expenses of 12% of precision. This result suggests that the training criterion used by the neural networks in the first step of our pipeline is more focused on precision than on recall. The last steps of our pipeline achieve a better balance

**TABLE 7**
**Effect of Data Augmentation on Recall.**

| | TRAIN | | | | TEST | | | |
|---|---|---|---|---|---|---|---|---|
| Model | WDA | DA | %Imp. | T | CNN | Pipeline | %Imp. | T |
| CNN1[a] | 1.00 (0.00) | 0.92 (0.03) | −0.08 | ≠ | 0.46 (0.02) | 0.57 (0.03) | +0.22 | ≠ |
| CNN1[b] | 1.00 (0.00) | 0.93 (0.05) | −0.07 | ≠ | 0.41 (0.04) | 0.60 (0.06) | +0.46 | ≠ |
| CNN2[a] | 1.00 (0.00) | 0.91 (0.03) | −0.09 | ≠ | 0.53 (0.03) | 0.67 (0.05) | +0.27 | ≠ |
| CNN3[a] | 1.00 (0.00) | 0.92 (0.02) | −0.08 | ≠ | 0.54 (0.03) | 0.69 (0.05) | +0.28 | ≠ |
| CNN3[b] | 1.00 (0.00) | 0.94 (0.02) | −0.06 | ≠ | 0.55 (0.04) | 0.70 (0.05) | +0.27 | ≠ |
| CNN4[a] | 0.99 (0.01) | 0.91 (0.04) | −0.08 | ≠ | 0.61 (0.05) | 0.70 (0.06) | +0.16 | ≠ |
| CNN4[b] | 1.00 (0.00) | 0.95 (0.04) | −0.05 | ≠ | 0.57 (0.02) | 0.69 (0.09) | +0.21 | ≠ |
| CNN4[c] | 1.00 (0.00) | 0.97 (0.04) | −0.03 | ≠ | 0.77 (0.12) | 0.90 (0.08) | +0.17 | ≠ |
| CNN5[a] | 1.00 (0.00) | 0.97 (0.04) | −0.03 | ≠ | 0.85 (0.11) | 0.88 (0.07) | +0.04 | = |

Notes: Comparison of the (average) precision obtained by the pipeline trained with the data augmentation strategy (column DA) and without it (column WDA). Results were computed and formatted as in Table 5.

**TABLE 8**
**Recall of the CNN and the Full Pipeline.**

| | TRAIN | | | | TEST | | | |
|---|---|---|---|---|---|---|---|---|
| Model | CNN | Pipeline | %Imp. | T | CNN | Pipeline | %Imp. | T |
| CNN1[a] | 0.87 (0.05) | 0.96 (0.01) | +0.10 | ≠ | 0.60 (0.06) | 0.71 (0.09) | +0.18 | ≠ |
| CNN1[b] | 0.83 (0.07) | 0.94 (0.02) | +0.13 | ≠ | 0.58 (0.08) | 0.68 (0.10) | +0.18 | ≠ |
| CNN2[a] | 0.84 (0.05) | 0.94 (0.03) | +0.13 | ≠ | 0.65 (0.09) | 0.78 (0.08) | +0.20 | ≠ |
| CNN3[a] | 0.86 (0.05) | 0.94 (0.04) | +0.09 | ≠ | 0.69 (0.07) | 0.78 (0.05) | +0.12 | ≠ |
| CNN3[b] | 0.85 (0.03) | 0.93 (0.03) | +0.10 | ≠ | 0.68 (0.06) | 0.77 (0.03) | +0.12 | ≠ |
| CNN4[a] | 0.88 (0.09) | 0.93 (0.04) | +0.06 | ≠ | 0.70 (0.10) | 0.76 (0.04) | +0.07 | = |
| CNN4[b] | 0.88 (0.11) | 0.96 (0.02) | +0.09 | ≠ | 0.65 (0.16) | 0.72 (0.04) | +0.11 | = |
| CNN4[c] | 0.59 (0.19) | 0.92 (0.03) | +0.56 | ≠ | 0.44 (0.17) | 0.79 (0.09) | +0.80 | ≠ |
| CNN5[a] | 0.63 (0.14) | 0.91 (0.03) | +0.45 | ≠ | 0.49 (0.13) | 0.81 (0.09) | +0.66 | ≠ |

Notes: Average recall obtained by the entire pipeline compared to the recall obtained by the CNN. Results were computed and formatted as in Table 1.

between these two metrics among the pullbacks, reducing false negatives with a slight decrease in true positives.

### 3.5. Qualitative Evaluation

In order to qualitatively evaluate the results of the pipeline, a few exemplar cases were selected by an expert. In such way, the advantage of the method and the limitation on some challenging sequences can be assessed.

The selected results were extracted from the sequences belonging to the test set and not from the training set.

Fig. 7 provides examples that the expert has identified as good cases of detections. In the first case (top panel), the prediction score is qualitatively and quantitatively optimal. Indeed, we can observe that the sequence-based refinement of the CNN predictions, which includes Gaussian smoothing and decision threshold optimization, improves the classification

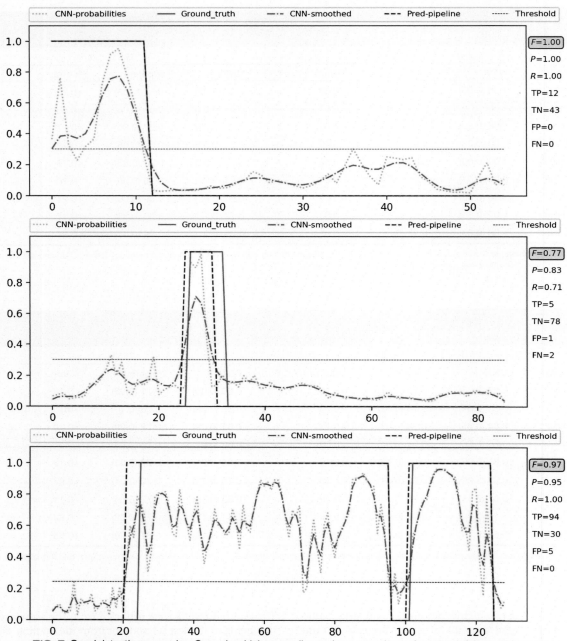

FIG. 7 Good detection examples. Cases in which, according to the expert, the performance of the method are satisfactory.

result. In particular, the Gaussian smoothing prevents the misclassification of frame 37 by averaging its value with the classification result of the neighboring frames. On the other hand, processing the CNN predictions with a canonical decision threshold $\theta = 0.5$ would have led to the detection of two calcification blocks instead of one (note the spike at frame 2) and errors in the boundaries. As a result, the postprocessing allows the system to correctly detect a single calcification block with precise boundaries. In this case, the $F$-measure score is equal to one, confirming the optimal detection of the sequence. The second case (mid panel) is interesting because it was selected as qualitatively good by the expert, but it exhibits a low $F_1$-score. This suggests that the classical detection metrics used to assess and rank the different models may be not accurate to indicate when a case is correctly detected under a clinical point of view. In this case, the system makes only one false positive (and two false negatives), but, as the number of positives is very small, this mistake translates into a sharp decrease in precision and $F_1$-score. In the last example (bottom panel), qualitative and quantitative evaluations match again: the system correctly identifies the number of calcification blocks and makes a slight mistake in the detection of the borders. In this case, the raw predictions of the CNN are visibly irregular and would lead to many spurious blocks, showing that the postprocessing phase is effective.

Fig. 8 provides examples that the expert has identified as average results. The system correctly predicts the number of calcification blocks but it is not very accurate in predicting the boundaries. In the most right block, the system is not able to compensate and correct the prediction obtained by the CNN because the classifier is highly confident about the absence of calcification after frame 60. In the middle block, the decision threshold optimization slightly decreases the detection of the right boundary of the calcium block. Indeed, in the second example (mid panel), the raw CNN detections would have been slightly more accurate in detecting the left boundaries of the calcification block perhaps at the expense of introducing a spurious block around frame 10. Instead the Gaussian smoothing makes both regions uniform, removing the sharp boundaries. In the last clinical case, the CNN predicts three calcification blocks, a central one and two short ones on the left and on the right of the sequence. The smoothing is effective and corrects the false block appear between frames 70 and 80 since the duration of the subsequence having positive CNN predictions is short. However, at the beginning of the pullback, the CNN is confident about

the presence of calcification in a longest subsequence of frames, and unfortunately the smoothing is not sufficient to eliminate the spurious block.

Fig. 9 presents cases that the expert has identified as bad (unsatisfactory) detections. In the first and second case (top- and mid-panels), the CNN predictions include many spikes (short and sharp changes) with many isolated false positives and false negatives. Such classification lead to detection of multiple short blocks of calcification, while in reality the pullback is composed of a few, long blocks of calcified plaque. The smoothing is not able to repair these cases since the Gaussian kernel used in the smoothing is too narrow. The dataset contains pullbacks that are very variable in length and the parameters of the filter were fixed to improve the *average* performance among all the training sequences. This problem may be overcome by using more powerful sequence refinement methods that, for instance, take into account the features of the specific sequence to determine the required degree of smoothing. Interestingly, these two cases corresponds to high $F_1$-scores, which suggests again that the classical machine learning metrics used to optimize the detection (precision and recall) may not be the optimal features that a clinician would use for considering a detection clinically satisfactory. The last case (bottom panel) is an example of the case in which the base detector (CNN) is so confident about its predictions that the system (based on these predicted likelihoods only) cannot repair the mistakes. In this case, the metrics correctly reflect the qualitative evaluation of the predictions.

## 4. CONCLUSIONS
We have presented a framework based on convolutional nets for the automatic selection of frames in an IVUS pullback that are relevant for the presence of arterial calcification. This can help a physician to focus on a small subset of the large sequence of frames produced by a typical IVUS data acquisition. Our best results were obtained with a deep model of five convolutional layers which outperformed the baseline and shallower CNNs, obtaining an average $F_1$-measure of 0.67, precision of 0.77, and recall of 0.83. Overall, our experiments suggest that the use of deep architectures for IVUS data analysis is promising even though challenging, due to the scarcity of labeled data. Careful architecting, data augmentation, and regularization are essential to obtain good results.

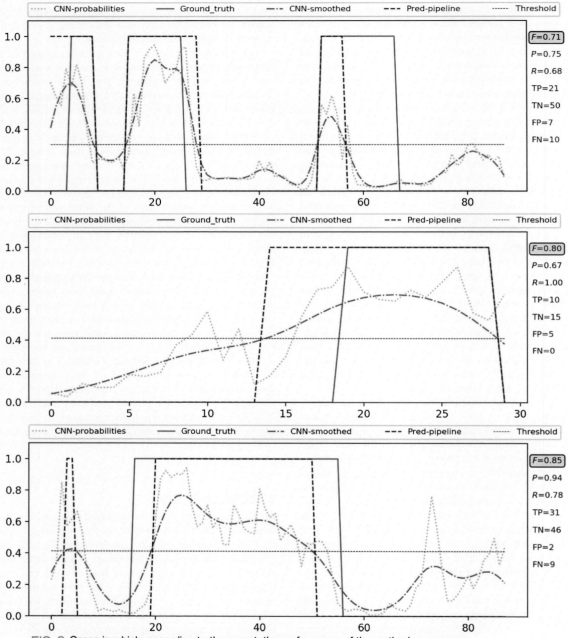

FIG. 8 Cases in which, according to the expert, the performance of the method are average.

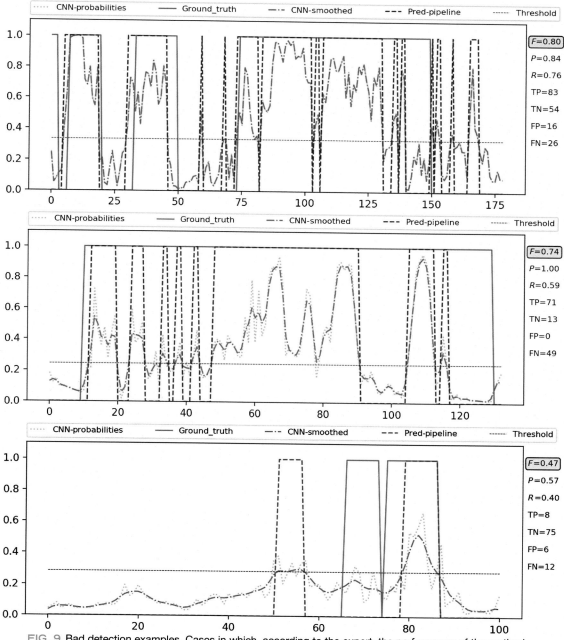

FIG. 9 Bad detection examples. Cases in which, according to the expert, the performance of the method are bad.

## ACKNOWLEDGMENT

This work was supported in part by the MICINN Grant RTI2018-095232-B-C21, 2017 SGR 1742, CERCA and ICREA Academia 2014.

## REFERENCES

1. Ciompi F, Pujol O, Gatta C, et al. Holimab: a holistic approach for media-adventitia border detection in intravascular ultrasound. *Med Image Anal.* 2012; 16 (6):1085–1100.
2. Alberti M, Balocco S, Gatta C, et al. Automatic bifurcation detection in coronary IVUS sequences. *IEEE Trans Biomed Eng.* 2012; 59(4):1022–1031.
3. Chen Z, Wahle A, Downe R, et al. Quantitative comparison of plaque in coronary culprit and non-culprit vessel segments using intravascular ultrasound. *Proceedings of MICCAI CVII-STENT Workshop, Boston, USA.* 2014:28–35.
4. Balocco S, Ciompi F, Rigla J, Carrillo X, Mauri J, Radeva P. Intra-coronary stent localization in intravascular ultrasound sequences, a preliminary study. *Intravascular Imaging and Computer Assisted Stenting, and Large-Scale Annotation of Biomedical Data and Expert Label Synthesis.* Springer; 2017:12–19.
5. Ciompi F, Pujol O, Balocco S, Carrillo X, Mauri J, Radeva P. Automatic key frames detection in intravascular ultrasound sequences. *Proceedings of MICCAI Workshop in Computing and Visualization for (Intra) Vascular Imaging (CVII), Houston, USA.* 2011:78–94.
6. Ciompi F, Balocco S, Rigla J, Carrillo X, Mauri J, Radeva P. Computer-aided detection of intracoronary stent in intravascular ultrasound sequences. *Med Phys.* 2016; 43 (10):5616–5625.
7. Ciompi F. Characterization in Intravascular Ultrasound [Ph.D. thesis]. Barcelona: University of Barcelona; 2012.
8. Gatta C, Balocco S, Ciompi F, Hemetsberger R, Leor OR, Radeva P. Real-time gating of IVUS sequences based on motion blur analysis: method and quantitative validation. *International Conference on Medical Image Computing and Computer-Assisted Intervention.* Springer; 2010:59–67.
9. LeCun Y, Boser BE, Denker JS, et al. Handwritten digit recognition with a back-propagation network. *Advances in Neural Information Processing Systems.* 1990:396–404.
10. Srivastava N, Hinton G, Krizhevsky A, Sutskever I, Salakhutdinov R. Dropout: a simple way to prevent neural networks from overfitting. *J Mach Learn Res.* 2014; 15(1):1929–1958.
11. Kingma DP, Ba J. Adam: a method for stochastic optimization. *arXiv preprint arXiv:1412.6980.* 2014.
12. Lipton ZC, Elkan C, Naryanaswamy B. Optimal thresholding of classifiers to maximize F1 measure. *Joint European Conference on Machine Learning and Knowledge Discovery in Databases.* Springer; 2014:225–239.
13. Joachims T. *Learning to Classify Text Using Support Vector Machines: Methods, Theory and Algorithms.* vol. 186. Norwell, MA: Kluwer Academic Publishers; 2002.
14. Nan Y, Chai KM, Lee WS, Chieu HL. Optimizing F-measure: a tale of two approaches. *arXiv preprint arXiv:1206.4625.* 2012.

# CHAPTER 10

# Computer-Aided Detection of Intracoronary Stent Location and Extension in Intravascular Ultrasound Sequences

SIMONE BALOCCO[a,b] • FRANCESCO CIOMPI[c] • JUAN RIGLA[d] • XAVIER CARRILLO[e] • JOSEPA MAURI[e] • PETIA RADEVA[a,b]

[a]Department of Mathematics and Informatics, University of Barcelona, Barcelona, Spain, [b]Computer Vision Center, Bellaterra, Spain, [c]Diagnostic Image Analysis Group, Pathology Department, Radboud University Medical Center, Nijmegen, The Netherlands, [d]InspireMD, Boston, MA, United States, [e]University Hospital Germans Trias i Pujol, Badalona, Spain

## 1. INTRODUCTION

An intraluminal coronary stent is a metal mesh tube deployed in a stenotic artery during percutaneous coronary intervention (PCI), in order to prevent acute vessel occlusion, to scaffold the arterial wall after balloon angioplasty, and to restore the blood flow. After X-ray-guided stent placement, cases of underexpansion (stent correctly apposed to luminal wall but not completely expanded) or malapposition (stent not completely in contact with the luminal wall) may happen. The incorrect stent deployment due to underexpansion and malapposition may lead to restenosis.[1] These clinical cases are recognized as important risk factors and might lead to stent failure (in-stent restenosis or stent thrombosis) in the follow-up.[2] The identification of the strut location and the definition of the stent shape, compared with the luminal border and the vessel border, allow physicians to assess stent placement in the vessel and the need for a further balloon postdilatation.

Moreover, the assessment of the stent location and extension along the vessel axis is relevant for PCI planning, implantation, and patient follow-up. In order to have an effective deployment, a stent should be optimally placed with regard to anatomical structures such as bifurcations and stenoses. The deployment of a stent in an incorrect location may lead to restenosis[1] or bifurcation side branch occlusion as shown by Ref. 3.

Although the current reference image modality for verifying the correct positioning of a stent is intravascular optical coherence tomography (OCT), a potential alternative is intravascular ultrasound (IVUS). IVUS is a catheter-based imaging technique that provides the sequence of tomographic images (pullback) of the internal vessel morphology (see Fig. 1A). The stent placement can be deduced by the position of its struts (see Fig. 1B). The main advantage of OCT over IVUS is that the resolution is 10-fold higher, easing the visualization of the stent for the physician. However, OCT remains inferior to IVUS in matters of penetration depth (1.5 vs. 5 mm),[4] which limits its ability to assess plaque burden, bifurcation angles,[5] and vessel remodeling. Moreover, the guiding OCT catheter requires the intubation of the coronary ostium in order to effectively provide contrast injection, limiting the analysis of ostial and occlusive lesions. Consequently, IVUS is regarded as either complementary or preferable to OCT in complex lesions.[6–8]

In this chapter, two complementary applications of the stent assessment in IVUS pullbacks are presented. The first technique, presented in Section 2.2, focuses on strut detection and stent shape estimation in a single IVUS image. The second method, presented in Section 3.2, uses the technique described on the struts and stent estimation to assess the position of the stent in the pullback.

## 2. STENT SHAPE ESTIMATION
### 2.1. State of the Art of Stent Shape Estimation

In IVUS images, often only few struts are visible in the IVUS image, due to the inclination of the ultrasonic

Intravascular Ultrasound. https://doi.org/10.1016/B978-0-12-818833-0.00010-2

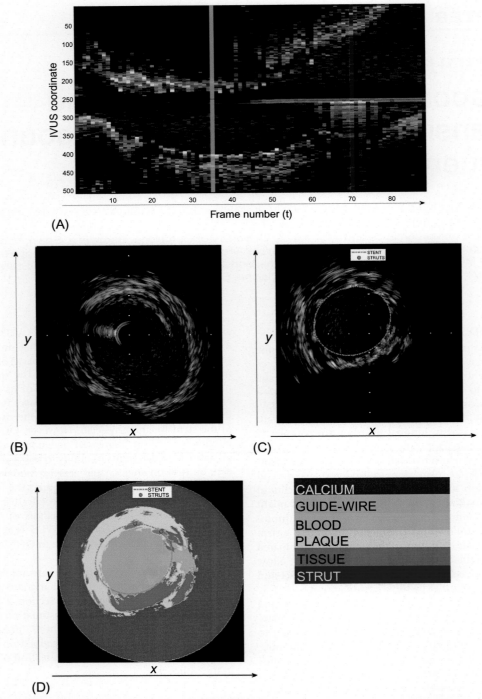

FIG. 1 Example of IVUS pullback containing a stent (A). In (B) and (C), two examples of IVUS frames in Cartesian coordinates, and belonging to the pullback (A), are depicted. The frame (B) corresponds to the *green vertical line* in (A), and does not contain a stent, while (C) corresponds to the *blue vertical line* in (A), and contains a stent. The *green circles* and the *green dotted line* in (C) and (D) represent the estimated position of struts and stent shape obtained using the technique of Ref. 9. The classification map (D) of the IVUS frame (C) is used to guide the struts and stent estimation. The legend of figure (D) indicates the tissue type identified in the image.

probe with respect to the longitudinal axis of the vessel and due to the presence of calcification or a dense fibrosis in contact with the stent. On the one other hand, several regions in the IVUS image may be confused with a strut, due to their local appearance. Clear examples of these regions are artifacts produced by the guidewire, refractions of ultrasonic waves, reverberations, or presence of small calcifications (see Fig. 2C and D). Additionally, strut appearance and thinning may vary significantly depending on the type of implanted stent, which can be either metallic or bioabsorbable.[10,11]

To date, several approaches for automatic stent analysis on OCT images have been presented.[12–14] However, only few approaches have been proposed in the field of IVUS image analysis,[15–19] which address either the problem of *stent modeling* or *strut detection*.

The problem of *stent modeling* was tackled by Ref. 16, where two deformable cylinders corresponding to the luminal wall and the stent were used. The cylinders were adapted to image edges and ridges to obtain a three-dimensional (3D) reconstruction of the boundaries. A semiautomatic stent contour detection algorithm was presented in Ref. 17. The method performs a global modeling of struts by minimum cost algorithm, followed by refinement using local information on stent shape and image intensity gradient. User interaction is finally foreseen to correct the stent shape in 3D. In Ref. 15, the same authors have proposed an improved version of this method, where the stent shape is accurately reconstructed in images with good quality, but the algorithm requires at least three clearly visible struts.

The problem of strut detection was instead tackled in Refs. 18, 19. An automatic method, limited to bioabsorbable scaffolds, was proposed by Ref. 18, using Haar-like features and a cascade of classifiers. A detection algorithm based on two-stage classification for fully automatic stent detection has been presented.[19] Despite the encouraging

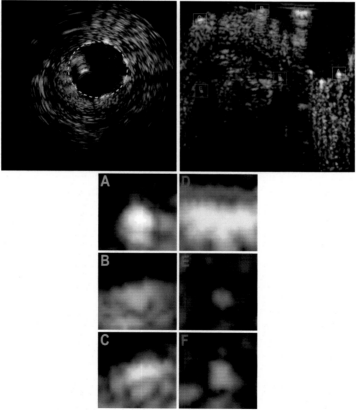

FIG. 2 Example of IVUS frame depicted in Cartesian (a) and polar (b) coordinates. In (c), details of the six regions highlighted in (b) are provided, for regions containing struts (A–C) and regions not containing struts (D–F), but visually similar to struts.

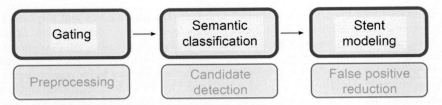

FIG. 3 Schematic representation of the proposed CAD system for stent analysis.

results, the number of false positive struts makes the method unsuitable for clinical purposes.

Finally, a novel CAD system for the fully automatic analysis of stents in IVUS image sequences is presented in Ref. 9. The overall layout of the methodology is depicted in Fig. 3. The system allows (1) to detect stent struts and (2) to provide an estimation of the stent shape, by considering contemporaneously the presence of struts and vessel morphology.

A curve approximating the stent shape is first estimated through a comprehensive interpretation of the vessel geometry. For this purpose, a semantic classification method is applied to each IVUS image, with the aim of providing a robust and accurate labeling of the vessel regions of interest in the frame. In the classification problem, the class *strut* is considered as one of the classes. For semantic classification purposes, the framework presented by Ref. 20 was adopted, tailoring the features to the specific stent-related problem. As a result, an optimal descriptor of the local appearance of tissues in the IVUS image is obtained. Successively, the struts are detected based on both local *appearance* and stent model, which allows a reduction in false positives.

The CAD was validated over a set of 107 IVUS sequences of in vivo arteries through a multicentric collaboration across five centers worldwide. The method has been validated on a dataset of 1015 IVUS images from the collected sequences, containing metallic stents and bioabsorbable ones.

## 2.2. Method
### 2.2.1. Gating
Let us define an IVUS pullback as a sequence of frames $I = \{f_i\}$. In the proposed pipeline, we first preprocess the pullback by applying an image-based gating procedure. Gating is a necessary step in order to make the analysis robust to two kinds of artifacts generated by the heart beating: the *swinging* (repetitive oscillations of the catheter along the axis of the vessel) and the *roto-pulsation* (irregular displacement of the catheter along the direction perpendicular to the axis of the vessel) effects. These two effects hinder a reliable identification of the luminal area,[21] which is an important factor in the analysis of

stent inside the vessel. For this purpose, the method presented by Ref. 22 is applied to the IVUS pullback, which selects a sequence of gated frames $G = \{f_{g_i}\}$ that are processed by the proposed CAD system (Fig. 4). It is worth noting that $G \subset I$ and that each index $g_j \approx 35i$, assuming a heart rate of approximately 60–80 bpm.

### 2.2.2. Semantic classification
In Ref. 9, the stent shape modeling is based on a detailed description of the vessel morphology, assessed by a frame-wise analysis of the pullback. For this purpose, semantic classification is applied to each frame in G, in order to obtain a pixel-wise labeling of each gated frame, taking into account the relationships between regions of the IVUS frame. In the following sections, we define (1) the classes considered in the semantic frame labeling, (2) the pixel-wise descriptor of local IVUS frame appearance, and (3) the framework used for semantic classification.

**Classes definition.** We consider that the semantic interpretation of an IVUS frame in the context of stent analysis requires the definition of the following six areas (see Fig. 5C): blood (B), plaque (P), calcium (C), guidewire (G), strut (S), and tissue (T), where it is important to stress that the strut areas indicate pixel candidates for containing a strut. In order to define these regions of interest, we rely on manual annotations of the following

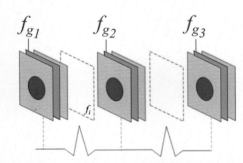

FIG. 4 Example of IVUS frames *f* and gated frames *g* in an IVUS pullback.

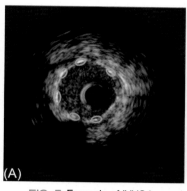 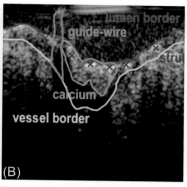 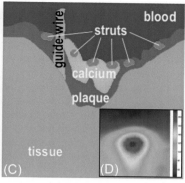

FIG. 5 Example of IVUS image in Cartesian coordinates where struts are indicated (A). In (B), the polar view of the image is depicted, and the main areas of interest are annotated. In (C), the semantic classes are indicated for the considered IVUS frame, obtained by processing the manual annotations in (B). In (D), the *strut filter* learned based on manual annotations is depicted.

contours in a set of training images (see Fig. 5B): lumen border, vessel border, calcified plaques, stent struts, and guidewire artifact, from which we derived the described areas of interested. For struts annotation purposes, we rely on a manually annotated set of points with coordinates $(x_s, y_s)$, which represent the center of struts, and consider the pixels belonging to a connected component area that includes the pixel in $(x_s, y_s)$ as belonging to the strut.

**Pixel-wise descriptor.** A descriptor of the local appearance in small neighborhoods of the IVUS frame is designed and associated to the central pixel of the neighborhood. Following a strategy proposed in several IVUS applications,[5,20] a combination of several feature types is used. The operators used to design the pixel-wise descriptor are partially derived from the optimal set of features defined by Ref. 20, namely, Gabor filters (GF), local binary patterns (LBP), Laplacian of Gaussian (LoG), normalized cross-correlation (NCC), shadow (SH), and matched filter response (MFR).

For Gabor filters, we adopted the set of parameters used by Ref. 20, obtaining a bank of 20 filters. The feature maps obtained from the convolution of the IVUS frame and each of the 18 filters are used to define a pixel-wise feature vector $\mathbf{x}_{GF} \in \mathbb{R}^{20}$.

The contribution of local binary pattern to the pixel-wise IVUS descriptor consists of the output of four configurations of the texture descriptor, using the pairs of values $(R, P) = ((1, 8), (2, 16), (3, 24), (4, 32))$. As a result, a feature vector $\mathbf{x}_{LBP} \in \mathbb{R}^4$ is obtained.

Stent struts appear as rounded-like objects in IVUS images, which can be interpreted as bright blobs. For this reason, we consider the response of the convolution with LoG operator at scales $\{37.5, 56.3, 75, 93.7, 112\} \mu$m, in order to cope with different strut size and imaging parameter settings, obtaining a feature vector $\mathbf{x}_{LoG} \in \mathbb{R}^5$.

In Refs. 23, 24, it has been shown that the NCC between patches of gated IVUS frames and adjacent frames in the pullback provides a low response in the presence of blood (see Ref. 23 for more details). Since the definition of the luminal area is an important factor in the analysis of stent, we adopt NCC as part of the pixel-wise descriptor. For this purpose, given a gated IVUS image $f_{g_j}$ of the sequence, we consider the two adjacent (nongated) frames $f_{g_j-1}$ and $f_{g_j+1}$ (see Fig. 5). Three measures of NCC are computed over three pairs of frames considering regions of size $W \times W$, namely $\text{NCC}_{g_j, f_{g_j+1}}$, $\text{NCC}_{g_j, f_{g_j-1}}$, $\text{NCC}_{f_{g_j-1}, f_{g_j+1}}$, and the average of the three responses is used as feature. It is worth noting that this descriptor takes advantage of local contextual information along the IVUS pullback. We apply this operator by varying $W$ from 0.1 to 0.2 mm with a step of 0.033 mm, and also consider the average across all the scales as an additional feature, obtaining a feature vector $\mathbf{x}_{NCC} \in \mathbb{R}^5$.

The presence of shadows in the IVUS image is an indicator of the presence of tissues or materials with high reflectance of ultrasonic waves, such as calcifications or stent struts. For this reason, we include in the descriptor two features named as *shadow* and *relative shadow*, extracted from the cumulative intensity of the ultrasound image along the radial direction (see Ref. 20 for more details), which form a feature vector $\mathbf{x}_{SH} \in \mathbb{R}^2$.

The last feature used in the proposed descriptor is based on the response of the IVUS frame to a learned representation of the strut appearance, which we call the *strut filter*. The appearance of the strut filter is learned from training examples. Given the set of annotated struts, we computed the filter by averaging the intensity of the gray level in a bounding box constructed around each position $(x_s, y_s)$. The size of the bounding box is defined as two times the diameter of the strut, estimated from the binary strut map obtained via Otsu thresholding of the gray levels of a sufficiently large region including struts in the training set. The obtained filter is depicted in Fig. 5D. The convolution of an IVUS frame and the strut filter gives high response in the presence of structures resembling struts. We use the pixel-wise value of the response as a feature of the descriptor, along with the pixel-wise product between the response and the IVUS image. In this way, we obtain a feature vector $\mathbf{x}_{SKR} \in \mathbb{R}^2$.

Finally, the IVUS image itself as well as its smoothed version are included in the descriptor, obtaining a descriptor $\mathbf{x}_{IVUS} \in \mathbb{R}^{40}$.

**Semantic classification framework.** The Multiscale Multiclass Stacked Sequential Learning (M$^2$SSL) approach is used for the purpose of multiclass semantic classification of IVUS frames, since it has been shown to provide a robust interpretation of IVUS images.[20,25] The M$^2$SSL is a classification architecture based on the stack of two classifiers, which we refer to as $H_1$ and $H_2$ (see Fig. 6A). The classifier $H_1$ is fed with the feature vector $\mathbf{x}_{IVUS}$ computed for each position $\vec{q} = (\rho, \theta)$ of the IVUS image in polar coordinates, and provides as output a vector $\mathcal{P} \in \mathbb{R}^{1 \times N_c}$ of pseudo-likelihoods, where $N_c$ is the number of classes. As in Ref. 20, the Error-Correcting Output Codes (ECOC) technique is used to deal with the multiclass problem and to compute the vector $\mathcal{P}(\vec{q})$ for any location of the image.

The classifier $H_2$ is fed with a combination of $\mathbf{x}_{IVUS}$ with features of *context* $\mathbf{x}_C$. The role of $\mathbf{x}_C$ is to encode long-range interactions between regions in the image, in order to explicitly encode semantic relationships between tissues. This is done through a functional $J(\mathcal{P})$, which performs multiscale sampling of the likelihood map $\mathcal{P}$ at $N_s$ scales (see Fig. 6A). At each scale $s$, the map is smoothed with a Gaussian kernel of standard deviation $\sigma_s = 2^{(s-1)}$ and then sampled in positions corresponding to the 8N neighborhood of each location $\vec{q}$, for each class (see Fig. 6A). The central pixel is included as well; as a consequence, $|\mathbf{x}_C| = 9 N_c N_s$. Finally, an *extended* feature vector $\mathbf{x}_E = \mathbf{x}_{IVUS} \cup \mathbf{x}_C$ is provided to $H_2$, which assigns the label $Y$ pixel-wise. In this way, the classifier $H_2$ is trained using both information on the local tissue appearance and information on semantic relationships among regions in the IVUS image. The result of the semantic classification, that is, the

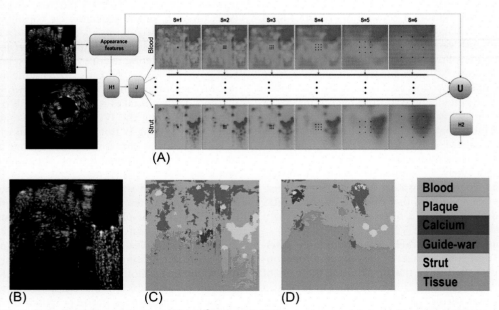

**FIG. 6** Example of the architecture of the M$^2$SSL classification framework (A). An example of IVUS image in polar coordinates is depicted in (B), along with the corresponding classification maps for the stages $H_1$ (C) and $H_2$ (D).

classification obtained after applying $H_2$, has less noise from a semantical point of view with respect to the result obtained only using $H_1$ (see Fig. 6C and D).

### 2.2.3. Stent modeling

In this section, we describe the procedure to assess the stent shape estimation and strut detection.

**Stent shape estimation.** Estimation the stent shape consists in finding a curve that simultaneously fulfills three criteria:

1. The shape should cross as many struts as possible.
2. The shape should fulfill morphological constraints with respect to the vessel structure (e.g., no struts inside a calcified region).
3. Given the rigidity of the stent meshes, the shape should be as regular as possible.

Considering (C), we assume the *ellipse* as a model for the stent shape, which has more degrees of freedom than the circular shape used in Ref. 15. In polar coordinates, the points of an ellipse are defined by the function $\rho(\theta)$, where $\rho$ is the distance from the ultrasound transducer and $\theta$ is the angle from the initial position of the transducer in the IVUS probe. The function can be written in a compact form as $\rho(\theta) = \frac{P(\theta) + Q(\theta)}{R(\theta)}$, with

$$\begin{cases} P(\theta) = \rho_0[(b^2 - a^2)\cos(\theta + \theta_0 - 2\phi) + (a^2 + b^2)\cos(\theta - \theta_0)], \\ Q(\theta) = \sqrt{2}ab\sqrt{R(\theta) - 2\rho_0^2\sin^2(\theta - \theta_0)}, \\ R(\theta) = (b^2 - a^2)\cos(2\theta - 2\phi) + a^2 + b^2, \end{cases} \quad (1)$$

where the set of parameters $\mathcal{E} = [a, b, \phi, \rho_0, \theta_0]$ indicates, in order, the major ($a$) and minor ($b$) axes, the orientation ($\phi$), and the coordinates ($\rho_0, \theta_0$) of the center of the ellipse. Following criteria (1, 2), a robust estimation of stent shape is obtained through a comprehensive interpretation of the curve position with respect to the vessel morphology. For this purpose, similar to Ref. 20, we take advantage of the output of the semantic classification by designing a functional $\Psi(\mathcal{E})$ that encodes the dependencies between parameters of the stent curve and vessel morphology:

$$\Psi(\mathcal{E}) = \sum_{i=1}^{|\vec{t}^+|} w_i t(\mathcal{E})_i^+ - \sum_{j=1}^{|\vec{t}^-|} w_j t(\mathcal{E})_j^-. \quad (2)$$

Based on the classification output, $\vec{t}^+ = \{t(\mathcal{E})_i^+\}$ and $\vec{t}^- = \{t(\mathcal{E})_j^-\}$ are the regions of the image positively or negatively contributing to the correct curve placement. Based on the nomenclature defined in Section 2.2.2, we define $\vec{t}^+ = \{B_\uparrow, P_\uparrow, C_\downarrow, B_\downarrow, P_\downarrow, C_\frown, P_\frown, S_\frown\}$ and $\vec{t}^- = \{C_\uparrow, S_\uparrow, T_\uparrow, S_\downarrow\}$, where the arrows indicate a region placed above ($\uparrow$), below ($\downarrow$), or crossed

($\frown$) by the curve. The shape is initialized by approximating the border of the convex hull containing the *blood* area of the IVUS image with an elliptical shape.

**Strut detection.** Once the stent shape is estimated, the positions $s_A = (\rho_A, \theta_A)$ representative for struts can be automatically detected by considering three conditions. First, the appearance of a strut is accounted by considering a likelihood map $\mathcal{L}_{strut}(\rho, \theta)$, obtained through the NCC between the IVUS image and the *filter* $K_S$ described in Section 2.2.2 (Fig. 5D). A local maximum of $\mathcal{L}_{strut}$ identifies a potential strut position. Strut detection is then subject to two additional conditions: (1) $s_A$ must belong to a region classified as strut $Q_{strut} = \{\vec{q}\}|_{Y(\vec{q})=strut}$; and (2) $s_A$ must be proximal with respect to the estimated stent shape. The coordinates of struts are then obtained as

$$\begin{cases} (\vec{\rho}_{strut}, \vec{\theta}_{strut}) = argmax_{\vec{\rho}, \vec{\theta}} (\mathcal{L}_{strut}(\vec{\rho}, \vec{\theta}) \cap Q_{strut}), \\ s.t. \\ dist((\vec{\rho}_{strut}, \vec{\theta}_{strut}), (\vec{\rho}_{stent}, \vec{\theta}_{stent})) \leq d_{strut}. \end{cases} \quad (3)$$

We used the value $d_{strut} = 0.2$ mm, which is the distance used in clinical practice to assess malapposition. As a result, a set of coordinates $S_A(f) = \{s_A\}$ are obtained for each frame $f$, representing the positions of the center of detected struts.

The pipeline of the proposed strategy for assessing the length and extension of intracoronary stent is depicted in Fig. 9. The three steps of the framework are detailed in the following sections.

### 2.3. Validation
### 2.3.1. Material

A set of 107 sequences of IVUS images was collected through a multicenter study, containing both *metallic* (*met*) and *bioabsorbable* (*abs*) stents. Data acquisition protocol was approved by the IRB of each clinical center. The IVUS sequences were acquired using iLab echograph (Boston Scientific, Fremont, CA) with a 40-MHz catheter (Atlantis SR Pro, Boston Scientific); no standardization of the echograph parameters was applied during the acquisitions. The pullback speed was 0.5 mm/s. The validation was performed for the two parts of the CAD system separately, the *frame-based* and the *pullback-based* analysis. Table 1 describes the datasets used in the evaluation process. The model of the metallic and bioabsorbable stents were "Promus Coronary Stent, Boston Scientific, MN" and "Absorb Bioresorbable Vascular Scaffold (BVS), Abbott, IL," respectively.

We collected 1015 frames from the 107 pullbacks. An expert manually annotated the beginning and the

**TABLE 1**
**Detailed Description of the Datasets Used in This Study.**

| Dataset type | Hospital | PULLBACKS | | N frames | LABEL | | |
| | | Number | Stent deployment | | Stent frames (%) | Training | Test |
|---|---|---|---|---|---|---|---|
| Metallic | No. 1 | 13 | Deployed | 90 | Roughly 50 | $train_{H1}$ | |
| | | | Deployed | 90 | Roughly 50 | $train_{H2}$ | |
| | | 80 | Deployed | 589 | 100 | | $test_{met}$ |
| | No. 2 | 3 | Deployed | 45 | Roughly 50 | | $test_{met}$ |
| | No. 3 | 5 | Deployed | 75 | Roughly 50 | | $test_{met}$ |
| Bioabs. | | 1 | Deployed | 21 | Roughly 50 | | $test_{abs}$ |
| | No. 4 | 4 | Deployed | 84 | Roughly 50 | | $test_{abs}$ |
| | No. 5 | 1 | Deployed | 21 | Roughly 50 | | $test_{abs}$ |

end of the stent in each sequence; more than one annotation per pullback was allowed when several stents were implanted in subsequent segments of the same artery. We split the dataset into two subsets: one was used for *training* and the other for *testing* purpose. The training set consisted of 13 pullbacks, from which we randomly selected 180 frames. All IVUS frames of the training set contained only one guidewire. For each frame in the training set, an expert manually annotated the contour of the six areas defined in Section 2.2.2, from which regions used to train the semantic classification framework are defined (see Section 2.2.2 for more details). The test set was derived from the remaining 94 pullbacks. From the test sequences, 835 frames were randomly selected, 709 containing metallic stents ($test_{met}$) and 126 containing bioabsorbable stents ($test_{abs}$). In each dataset, the proportion of frames with and without stent was roughly 1:1, with the exception of the test set provided by the hospital No. 1, having only stent frames (see Table 1). In each frame, two observers were asked to independently annotate the locations of stent struts in Cartesian coordinates, which we converted into polar coordinates $(\rho_M^i, \theta_M^i)$.

### 2.3.2. Experiments on stent modeling

The detection of struts and modeling of the stent shape is based on the semantic classification of IVUS frames. In order to train $H_1$ and $H_2$ with different data, *train* was split into two balanced subsets $train_{H1}$ and $train_{H2}$, containing 90 frames each. In our experiments, Adaptive Boosting[26] was used for training both $H_1$ and $H_2$ classifiers. The multiclass problem was handled using the ECOC[27] framework with one-vs-one strategy to train binary dichotomies. As a result, $H_1$ and $H_2$ consisted of 15 binary classifiers each. The number of scales in M$^2$SSL was $N_s = 6$, which allows to encode long-range

interactions that cover up to half of the image size. The training of the weights $\mathbf{w} = \{w_i, w_j\}$ was done in cross-validation by approximating manual struts annotations with an elliptic model, and then averaging the normalized amount of tissues for each frame of the training set.

Struts were detected in position $(\rho_A^j, \theta_A^j)$ and the stent modeled as detailed in Sections 2.2.2 and 2.2.3. First, we compared the automatic detection with the manual annotation $(\rho_M^i, \theta_M^i)$ and computed true positives (TP), false positives (FP), and false negatives (FN). For evaluating the detection, we considered a strut detected at $(\rho_A^j, \theta_A^j)$ as being inside a circular region $\mathcal{N}$ of radius $\Phi$ around each $(\rho_M^i, \theta_M^i)$:

- if $(\rho_A^j, \theta_A^j) \in \mathcal{N} \Rightarrow$ TP;
- if $(\rho_A^j, \theta_A^j) \notin \mathcal{N} \Rightarrow$ FP; and
- if $\mathcal{N} \cap \{\rho_A^j, \theta_A^j\} = \varnothing \Rightarrow$ FN.

Based on these criteria, parameters of *Precision* (P), *Recall* (R), and *F-measure* (F) were computed.

Second, we evaluated the quality of the information provided by the detected strut to estimate the stent shape. For this purpose, we considered the distance from the automatically computed curve to the manual struts. The performance was assessed in terms of both (a) radial distance $d_{SC}$ between the strut points $(\rho_M^i, \theta_M^i)$ and the stent curve $(\vec{\rho}_{stent}, \vec{\theta}_{stent})$ and (b) radial distance $d_{SS}$ between $(\rho_M^i, \theta_M^i)$ and $(\rho_A^j, \theta_A^j)$. The results in terms of both detection performance and distance are presented in Table 2 for metallic stents and in Table 3 for bioabsorbable stents. Visual results are depicted in Fig. 7 for metallic stent, where results are grouped by categories of frames containing: *small-large* vessels (columns A and B, rows 4 and 5), frames containing a *bifurcation* (columns C and D), and frames containing

**TABLE 2**
Quantitative Results (Average and Standard Deviation) Reporting the Performance of the Algorithm Over Dataset $test_{met}$.

| | | Precision (P) Mean (std) | Recall (R) Mean (std) | F-measure (F) Mean (std) | Strut to strut distance ($d_{SS}$) Mean (std) | Strut to curve distance ($d_{SC}$) Mean (std) |
|---|---|---|---|---|---|---|
| All | Auto vsobs-1 | 76.4% (27.5%) | 89.9% (19.6%) | 77.7% (24.2%) | 0.10 mm (0.04 mm) | 0.15 mm (0.12 mm) |
| | Auto vsobs-2 | 78.4% (28.6%) | 84.8% (22.7%) | 75.7% (25.7%) | 0.09 mm (0.04 mm) | 0.14 mm (0.11 mm) |
| | Obs-1 vsobs-2 | 86.6% (20.8%) | 93.3% (17.1%) | 86.9% (19.0%) | 0.07 mm (0.05 mm) | |
| Bifurcation | Auto vsobs-1 | 63.3% (32.9%) | 88.9% (20.6%) | 66.3% (28.9%) | 0.11 mm (0.04 mm) | 0.17 mm (0.12 mm) |
| | Auto vsobs-2 | 62.9% (33.6%) | 80.0% (25.4%) | 62.5% (29.6%) | 0.09 mm (0.04 mm) | 0.18 mm (0.14 mm) |
| | Obs-1 vsobs-2 | 80.0% (26.7%) | 92.8% (16.5%) | 82.0% (23.6%) | 0.07 mm (0.06 mm) | |
| Calcium | Auto vsobs-1 | 69.2% (29.3%) | 89.7% (21.1%) | 71.9% (26.3%) | 0.10 mm (0.04 mm) | 0.17 mm (0.13 mm) |
| | Auto vsobs-2 | 73.0% (29.8%) | 80.0% (26.1%) | 68.4% (25.3%) | 0.10 mm (0.04 mm) | 0.17 mm (0.13 mm) |
| | Obs-1 vsobs-2 | 82.1% (23.5%) | 93.5% (16.4%) | 84.0% (19.5%) | 0.07 mm (0.06 mm) | |
| Small-large | auto vsobs-1 | 54.0% (36.2%) | 86.7% (23.2%) | 57.3% (31.2%) | 0.11 mm (0.04 mm) | 0.21 mm (0.15 mm) |
| | Auto vsobs-2 | 55.5% (32.5%) | 82.6% (27.1%) | 60.6% (30.0%) | 0.10 mm (0.04 mm) | 0.23 mm (0.15 mm) |
| | Obs-1 vsobs-2 | 85.0% (23.1%) | 85.8% (27.0%) | 79.2% (24.1%) | 0.07 mm (0.05 mm) | |
| Normal | Auto vsobs-1 | 77.4% (26.6%) | 91.1% (18.7%) | 79.4% (21.8%) | 0.10 mm (0.04 mm) | 0.15 mm (0.12 mm) |
| | Auto vsobs-2 | 79.6% (27.3%) | 86.7% (21.9%) | 78.2% (22.9%) | 0.09 mm (0.04 mm) | 0.14 mm (0.11 mm) |
| | Obs-1 vsobs-2 | 87.6% (20.4%) | 94.0% (16.3%) | 88.1% (18.2%) | 0.07 mm (0.05 mm) | |

Notes: The results are separated according the following categories: all the frames (all), presence of bifurcations (bifurcation), calcification (calcium), small-large vessel (small-large), and absence of previous cases (normal). The mean and standard deviation (in parenthesis) distances between automatic and manual struts ($d_{SS}$), and between manual struts and stent curve ($d_{SC}$) are indicated.

calcium plaques (columns E and F). Frames not belonging to the previous categories (columns A and B, rows 1–3) were labeled as normal. In Fig. 8, visual results for cases of bioabsorbable stent are depicted. In this case, due to the fewer number of examples, solely two categories (all) and (stent-only) were considered.

## 2.4. Discussion
### 2.4.1. Metallic stents
When frames containing metallic stents are processed, an average performance of ($F_{obs-1} = 77.7\%$ and $F_{obs-2} = 75.7\%$ on all frames) is reached. The quantitative results

of this study outperform methods based on similar approaches (F-measure of 71% in Ref. 24, F 66% in Ref. 19), although the comparison is not straightforward since the performances reported in such studies were obtained over different datasets. The method performs well in the presence of normal frames as well as in the presence of bifurcations and calcified plaques (Table 2). This represents an interesting results, since in terms of manual annotations, a high interobserver variability score is obtained in the category bifurcation as well as in frames containing calcium where the two

**TABLE 3**
**Quantitative Results (Average and Standard Deviation) Reporting the Performance of the Algorithm Over Dataset *test_abs*.**

|  | Precision (*P*) Mean (std) | Recall (*R*) Mean (std) | F-measure (*F*) Mean (std) | Strut to strut distance (*d_SS*) Mean (std) | Strut to curve distance (*d_SC*) Mean (std) |
|---|---|---|---|---|---|
| *Auto* vs*obs-1* | 87.0% (22.8%) | 80.4% (24.5%) | 78.8% (22.8%) | 0.09 mm (0.04 mm) | 0.09 mm (0.07 mm) |
| *Auto* vs*obs-2* | 80.9% (27.6%) | 76.1% (22.4%) | 73.3% (23.3%) | 0.10 mm (0.04 mm) | 0.11 mm (0.08 mm) |
| *Obs-1* vs*obs-2* | 88.6% (18.5%) | 85.2% (22.5%) | 83.7% (17.5%) | 0.09 mm (0.04 mm) |  |

*Notes*: The results are separated according in two categories: all the frames (*all*) and frames containing struts (*stent-only*). The mean and standard deviation distances between automatic and manual struts ($d_{SS}$), and between manual struts and stent curve ($d_{SC}$) are indicated.

observers, in some cases, placed the struts in different location of the plaque. Lower performance is obtained on frames with small and large vessels, mainly due to a suboptimal initialization of the stent shape.

It can be noted that in the presence of normal and *small-large vessels* (Fig. 7A and B), most of the struts selected by the observers are identified by the automatic method (rows 1–3) and the estimated stent curve fits well all the manually labeled struts. However, when the vessel is narrow (stenotic) or large (Fig. 7, rows 4 and 5), we observed an underestimation of the lumen area, which affects the estimation of the stent shape.

In the presence of *bifurcations* (Fig. 7C and D), we can notice disagreement between the two observers (rows 4 and 5). Nevertheless, automatic detection of the proposed CAD system matches the annotation of at least one of the two observers. In Fig. 7D,5, two concentric stents are present in the vessel, the first one is covered by a plaque, and a second one deployed in the inner lumen area. As a result, the automatic stent contour fits the struts closer to the catheter.

In the presence of *calcifications* (Fig. 7E and F), the method is able to correctly detect struts even if located close to calcified plaques (rows 1–3). However, in other cases the close contact with a calcified plaque hampers the correct detection of struts (rows 4 and 5). As a consequence, the estimated stent shape does not follow the vessel boundary. In the IVUS frame in Fig. 7F,5, most of the struts are identified; however, in the top area the catheter shadow partially occupy the lumen area, preventing the stent shape to fit all the detected struts.

**Bioabsorbable stents.** In the presence of bioabsorbable stents, the performance of the strut detection

(Table 3) are comparable with the results obtained on the metallic frames ($F_{obs-1} = 78.8\%$ and $F_{obs-2} = 73.3\%$). The performance reported in this chapter are comparable with the numerical results provided by Ref. 18 (*F-measure* of 71% and 75%), although the comparison is not straightforward, since they were obtained over a different dataset. This result is interesting, since the algorithm for strut detection was only trained with the local appearance of the metallic stent, demonstrating that the method is flexible and can be applied to IVUS images containing a more heterogeneous variety of stent types.

Most of the struts selected by the observers are identified by the automatic method (see Fig. 8). It is interesting to note that a reasonable estimation of the stent shape is obtained even if few struts are present in the image (Fig. 8A, bottom). In Fig. 8 (B, bottom), the stent shape is slightly overestimated and the stent contour crosses the vessel border in the bottom part of the image. In Fig. 8 (C, top and bottom), cases of false positive and false negative detection are shown, respectively. In particular, in Fig. 8 (C, top), it is interesting to note that a strut on the bottom of the image is correctly identified even if located close to the guide shadow reverberations. Finally, two cases in which the detection performance of the algorithm are low are illustrated in Fig. 8 (column D). In Fig. 8 (D, top), the agreement between the observers is low, and the algorithm detect struts identified by one of the two observers. In Fig. 8 (D, bottom), the algorithm incorrectly identifies bright scatterers on the vessel border as struts.

Although the numerical performance obtained in this study outperform the qualitative results reported in previous approaches, a fair comparison of several

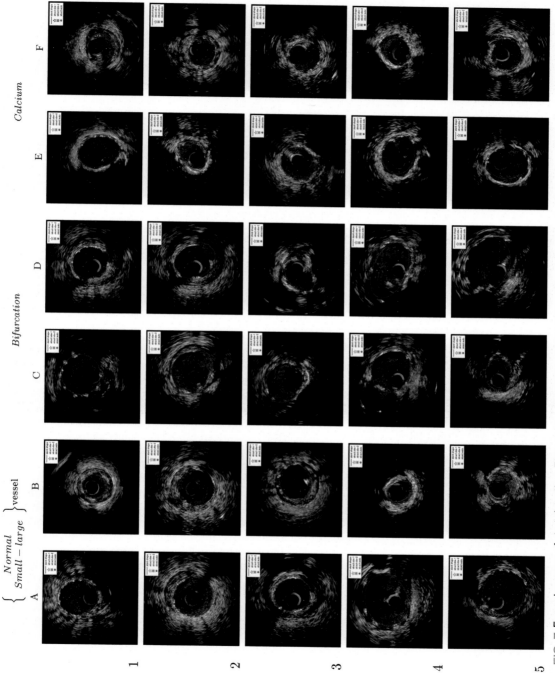

FIG. 7 Exemplar cases of strut detection. Frames are grouped according to the following categories: normal small and large vessel (A and B columns), bifurcation (C and D columns), and calcium (E and F columns). In each image are depicted two observer annotations (in *green squares* and *yellow star markers*) and the results of the automatic method (represented with *red circle markers*), respectively. The stent shape automatically computed is outlined using a *blue solid line*. (For interpretation of the references to color in this figure legend, the reader is referred to the web version of this article.)

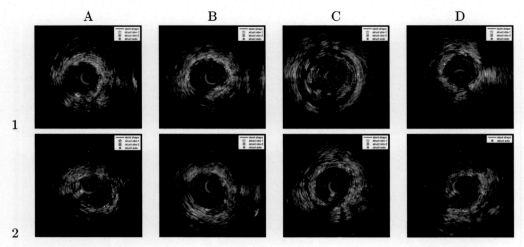

FIG. 8 Examples of bioabsorbable strut detection. Cases in which an accurate detection is obtained are depicted in columns (A, B), while in columns (C, D), cases with suboptimal results are considered. Manual annotation from two observers (obs-1, obs-2) is indicated as well as the results from the automatic method (auto). The stent shape estimated according to the method detailed in Section 2.2.3 is also depicted as a *continuous line* (best viewed in colors).

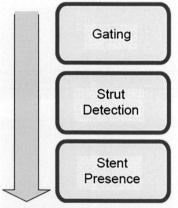

FIG. 9 Schematic representation of the proposed system for automatic analysis of stents.

methods over the same dataset would only be possible by competing in a challenge such as those presented in Ref. 28. The method has been implemented in Matlab (The MathWorks, Natick, MA, 2011) and the computation time is 0.33 s per frame, using a Intel i7 quad-core processor.

## 2.5. Conclusion

We have presented a framework for computer-aided detection (CAD) of intracoronary stents in IVUS image sequences to detect the struts and provide an estimation of the stent shape. The results obtained on an heterogeneous dataset containing both metallic and bioabsorbable stents, collected through a multicentric collaboration are close to the interobserver variability and suggest that the system has the potential of being used during percutaneous interventions.

The method achieved performance of strut detection with an overall F-measure of 77.7% with a mean distance of 0.15 mm from the manually annotated struts in the case of metallic stent, while for bioabsorbable stents an overall F-measure of 77.4% was obtained, with a mean distance of 0.09 mm from the manually annotated struts. Although the numerical performance obtained in this study outperform the qualitative results reported in previous approaches, a fair comparison of several methods over the same dataset would only be possible by competing in a challenge such as Ref. 28.

The implicit analysis of cases of malapposition and underexpansion is out of scope for this chapter and represents part of our future research. Such analysis could be performed, for instance, by combining the CAD framework described in this chapter with other tools.[20,23]

The image-base gating procedure is aimed at identifying the most stable frames of the sequence, which correspond to the end-diastolic phase. Such procedure allows an efficient computation of the NCC feature, along adjacent frames (not gated frames) as shown by Balocco et al. 23. However, the gating procedure does not guarantee that successive gating frames (and consequently the struts contained in the image) are aligned, so as to track individual struts. Such goal is planned for future research and can be obtained by applying the technique developed by Gatta et al. 29.

The system has been validated using cases of both metallic and bioabsorbable stents. It is worth noting that the system has been solely trained using data from cases with metallic stent. Bioabsorbable stents are used less often during PCIs, therefore finding a large set of examples to train a new system is not a trivial task. However, good performance can be observed also in the presence of bioabsorbable stents, showing the capability of the system to generalize well to detection of unseen examples, including other types of stents.

# 3. LONGITUDINAL STENT LOCATION
## 3.1. State of the Art of Longitudinal Stent Location

IVUS images can be also visualized in a long-axis view, allowing a pullback-wise analysis and in short-axis view for frame-wise analysis (see Fig. 1A and B). The physician examines both views in order to perform a diagnosis of the correct stent deployment. On one hand, the pullback-wise analysis allows the physician to localize the position of the stent within other vascular structures and produce a rough estimate of stent length. On the other hand, the frame-wise analysis allows a refinement of the assessment by detecting the initial and the final frame in which the stent is present, by assessing if struts appear in a frame. Some struts might not be visible in the long-axis view of the pullback because of the chosen angle for the visualization.

To date, several approaches allowing the detection and 3D reconstruction of stents from OCT images have been published,[12,30,31] since the high image resolution allows a detailed identification of the struts along the pullback. Instead, in IVUS, the visualization and rendering of the 3D shape of the stent is feasible,[32,33] but the accurate strut detection and the longitudinal localization of the stent along the pullback remains a challenging task.[34,35] In IVUS images, only a few struts are often visible, due to the inclination of the ultrasonic probe with regard to the longitudinal axis of the vessel. Several

regions in the IVUS image may look similar to a strut, due to their local appearance (guidewire artifacts, refractions of ultrasonic waves, reverberations, or presence of small calcifications or dense fibrotic tissue in contact with the stent) as illustrated in Fig. 1B and C. Additionally, the strut texture and the thickness may vary depending on the type of implanted stent, which can be either metallic or bioabsorbable.[10,11] As a consequence, the analysis of a single short-axis image may not be sufficient for accurately assessing if struts are present. In most of ambiguous cases, the physician has to scroll the pullback back and forward, analyzing adjacent frames in order to identify the frames defining the stent boundaries.

Few approaches for automatic stent analysis in IVUS have been proposed so far.[9,15–19] All the existing methods aim at identifying the struts in the short-axis view of the pullback assuming that the analyzed frame always contains a stent; these methods therefore rely on the presence of struts in the axial frame. Indeed, so far no strategies have been proposed for detecting if the stent is present in the frame, and for finding the boundaries and the position of the stent along the pullback. A possible reason could be that in most of the published techniques,[9,15–19] the number of false positive struts were too high, making it difficult to differentiate a frame containing a stent (a stent frame) from others in an IVUS pullback. Instead, this chapter extends a strut detection method for metallic and bioabsorbable stent that results more robust then previous stent detection approaches,[9] and using the strut detection obtained in successive short-axis views, estimates the location and extension of the stent along the pullback.

Finally, in Ref. 36, a novel approach that automatically detects the boundaries and the position of the stent along the IVUS pullback is presented.

The overall layout of the methodology of Ref. 36 is depicted in Fig. 9. The pipeline of the method consists in three steps: (1) the identification of the stable frames of the sequence, (2) the detection of stent struts, and (3) the assessment of the stent presence (location and extension) along the pullback. In this second pipeline, the image-based gating technique presented in Section 2.2.1 (introduced in Ref. 22) and a strut detection method described in Section 2.2 (introduced in Ref. 9) were used to obtain (1) and (2). Such techniques were chosen considering the superior performance with regard to other state-of-the-art options, while a novel approach has been designed for (3). In particular, phase (3) consists in defining a measure of likelihood for a frame to contain a stent, which we call *stent presence*. A temporal series is obtained by computing such likelihood along the whole sequence. The mono-dimensional signal is modeled as a train of rectangular waves by using an iterative and multiscale approximation of the signal to symbols using the SAX (Symbolic Aggregate approXimation) algorithm,[37] which was initially introduced as low computational complexity method for classic data mining tasks such as clustering, classification, and indexing. In this chapter, SAX is used to obtain an unsupervised and robust binary representation of the signal representing the *stent presence* in a pullback.

The technique was extensively validated using a set of 103 IVUS sequences of in vivo coronary arteries containing metallic and bioabsorbable stents, which are acquired through an international multicentric collaboration across five centers. Precision, Recall, F-measure, and Jaccard scores along with the distance between detected and reference frames are reported, and compared against interobserver and intraobserver measurements of two experts.

Additionally, application of the framework to each IVUS gated frame allows to estimate the lumen and stent contours. Using such information it is possible to compute the stent malapposition for each angular sector of the frame along the pullback. As a result, a map indicating the distance between the lumen and the stent in each angular section of the pullback is created and the angular sector of the pullback displaying malapposition is identified.

The pipeline for the CAD system for intracoronary stent analysis is depicted in Fig. 9. The three steps of the pipeline are detailed in the following sections.

## 3.2. Method
### 3.2.1. Gating and strut detection
As described in Section 2.2.1, a sequence of gated frames $G = \{f_{g_j}\}$ is obtained by applying the method presented by Gatta et al. 22 to the IVUS pullback.

The stent struts were detected by applying the strut detection method proposed by Ciompi et al. 9 and described in Section 2.2 to each gated frame independently. The method identifies the stent struts by contemporaneously considering the textural appearance of the stent and the vessel morphology. The strut detection system uses the Multiscale Multiclass Stacked Sequential Learning ($M^2SSL$) classification scheme to provide a comprehensive interpretation of the local structure of the vessel. In the classification problem, the class *strut* is considered as one of the six classes (defined as blood area, plaque, calcium, guidewire shadow, strut, and external tissues). For semantic classification purposes, tailored features used for classification of the problem[20] are used.

As a result, for each pixel $p(x, y)$ of a gated IVUS image, a classification map $M$ is obtained (see Fig. 1D). A curve approximating the stent shape $S_{shape}$ is initially estimated considering vascular constrains and classification results. For each region of $M$ labeled as stent $(M_{\{S\}})$, a strut candidate is considered. The selected struts $p_s(x, y)$ were selected among the candidates, considering both local *appearance* and distance with regard to the stent shape $S_{shape}$. Consequently, false positive candidates were discarded, and the regions containing a selected strut struts $M^*_{\{S\}}$ are a subset of $M_{\{S\}}$.

### 3.2.2. Stent presence assessment
The frames of the pullback corresponding to the vessel positions where the stent begins and ends can be identified by analyzing the detected struts. We model the presence of stent as a rectangular function $\sqcap(t)$, where the variable $t$ indicates the temporal position in the pullback (see Fig. 1). We estimate the binary signal $\sqcap(t)$ by processing a real-valued signal $\gamma(t)$, which we define as *stent presence*, corresponding to the frame-based likelihood of finding a stent in each frame of the IVUS sequence. The value of $\gamma(t)$ for each position $t$ in the sequence is computed by considering both the number of struts and their area, thus negatively weighting small strut areas of the images which have an high probability to be incorrect detections.

$$\gamma(t) = \sum_{p \in M^*_{\{S\}}} p|_{p_s \in M^*_{\{S\}}}, \qquad (4)$$

where $p_s \in M^*_{\{S\}}$ indicates the pixels of the IVUS frame labeled as *strut* containing an selected strut. An example of signal $\gamma(t)$ is depicted in Fig. 10C.

The signal $\gamma(t)$ may contain several transitions between low and high amplitudes, due to the variability in the number of struts visible in consecutive frames and due to suboptimal strut detection. For this reason, we filter the $\gamma(t)$ signal by considering its local statistics applying the SAX algorithm.[37] SAX is a symbolic representation algorithm that estimates a quantization of the time series based on global signal measurements and local statistics of subsequent neighbor samples. Given the signal $\gamma(t)$ and a window size $w$, the algorithm calculates a Piece-wise Aggregate Approximation (PAA) $\widehat{\gamma(t)}$, which is obtained by computing the local average values of $\gamma(t)$ over $n_w$ segments $w$-wide. Each average value is then normalized over the signal $\gamma(t)$. The procedure first computes a vector $\overline{\gamma(t)} = (\overline{q}_1, \dots \overline{q}_{n_w})$ where each of $\overline{q}_i$ is calculated as follows:

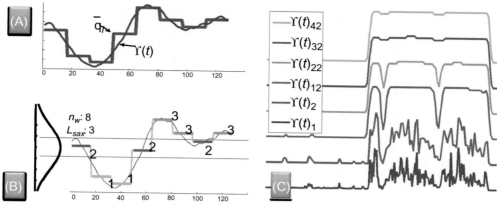

FIG. 10 In order to illustrate with an example the procedure of SAX algorithm, a Piece-wise Aggregate Approximation of a synthetic curve, not related with any measurement, is depicted in (A) and (B). The method first computes the local average values $\bar{q}_i$ of the signal $\gamma(t)$ over $n_w$ segment (A). Then, the algorithm normalizes the samples by the mean and standard deviation of the signal obtaining the amplitudes $\hat{q}_i$. Considering a Gaussian distribution of the samples amplitudes, the samples $\hat{q}_i$ are mapped to $L_{sax}$ discrete values having equi-probability (B), that is, the cutoff values are chosen in such way to separate the area below the Gaussian curve in $L_{sax}$ equal regions. In (A) and (B), the signal is quantized using $n_w = 8$ and $L_{sax} = 3$. Then, in (C), the SAX algorithm is applied to a signal $\gamma(t)$ obtained from an IVUS pullback. In this case, the signal $\gamma(t)$ is quantized using $n_w = 2$ and $L_{sax} = 2$. The SAX algorithm is iterated $N_{sax}$ times over an exemplar signal $\gamma(t)_i$, where $i$ is the iteration number. In case (C), the signals $\gamma(t)_i$ obtained in 6 of the 50 iterations (namely $i = 1, 2, 12, 22, 32, 42$) are illustrated.

$$\bar{q}_i = \frac{1}{w} \sum_{j=w(i-1)+1}^{w \cdot i} q_j, \tag{5}$$

where $i, j \in \mathbb{N}$. The quantified signal amplitudes $\hat{q}_i$ obtained by normalizing $\bar{q}_i$ by mean $\mu_\gamma$ and standard deviation $\sigma_\gamma$ of the signal $\gamma(t)$:

$$\hat{q}_i = \frac{\bar{q}_i - \mu_\gamma}{\sigma_\gamma}. \tag{6}$$

Then, the $\hat{q}_i$ amplitudes are mapped to $L_{sax}$ discrete values. Considering a Gaussian distribution of the samples amplitudes, each value of $L_{sax}$ corresponds to an equal-sized area of the Gaussian curve.

The SAX algorithm is iterated $N_{sax}$ times, until converging to flat intervals along the signal $\gamma(t)$. The maximum iteration number $N_{sax}$ is achieved when the difference between subsequent iterations of SAX is zero. Fig. 10C illustrates the iteration of the SAX algorithm over an exemplar signal $\gamma(t)$. The other parameter of the SAX algorithm is the number of quantized values assigned to the signal $L_{sax}$. The iterative SAX algorithm is described by the following equation:

$$\gamma(t)_{k+1} = SAX\left(\gamma(t)_k, \sigma_k^{train_{sax}}, \mu_k^{train_{sax}}, L^{train_{sax}}\right), \tag{7}$$

where $k \in 1 \ldots N_{sax}$ and $\sigma_k^{train_{sax}}$ and $\mu_k^{train_{sax}}$ are the mean and standard deviation computed on the training set at the iteration $k$, and the number of quantized values $L_{sax}$ is a constant that has been optimized by cross-validation using the training set, as described in Section 3.3. When the SAX algorithm reaches the maximum iteration number $N_{sax}$, the binary signal indicating the stent presence of the stent is obtained as $\sqcap(t) = \gamma(t)_{N_{sax}} > \mu_{N_{sax}}^{train_{sax}}$.

## 3.3. Validation
### 3.3.1. Material
A set of 103 sequences of IVUS images was collected through a multicenter study (see Table 4). In this study, we are interested in assessing the performance of the method at detecting the presence of stent along the pullback. For this reason, pullbacks acquired before and after stent deployment were used, therefore considering also sequences with the absence of stent. The dataset includes nine pullback acquired before the stent deployment. The remaining 94 pullbacks contain one or more *metallic* (*met*) or *bioabsorbable* (*abs*) stents.

Data acquisition protocol was approved by the IRB of each clinical center. The IVUS sequences were acquired using iLab echograph (Boston Scientific) with a 40-MHz catheter (Atlantis SR Pro, Boston Scientific);

**TABLE 4**
Detailed Description of the Datasets Used in This Study.

| Dataset type | Hospital | Number of pullbacks | Stent deployment | Stent frames (%) | Training | Test |
|---|---|---|---|---|---|---|
| Metallic | No. 1 | 80 | Deployed | Roughly 50 | $train_{met}$ | $test_{met}$ |
| | No. 2 | 3 | Deployed | Roughly 50 | $train_{met}$ | $test_{met}$ |
| | | 1 | Predeployed | 0 | $train_{met}$ | $test_{met}$ |
| | No. 3 | 5 | Deployed | Roughly 50 | $train_{met}$ | $test_{met}$ |
| | | 3 | Predeployed | 0 | $train_{met}$ | $test_{met}$ |
| Bioabsorbable | | 1 | Deployed | Roughly 50 | $train_{met}$/ $train_{abs}$ | $test_{abs}$ |
| | | 1 | Predeployed | 0 | $train_{met}$/ $train_{abs}$ | $test_{abs}$ |
| | No. 4 | 4 | Deployed | Roughly 50 | $train_{met}$/ $train_{abs}$ | $test_{abs}$ |
| | | 3 | Predeployed | 0 | $train_{met}$/ $train_{abs}$ | $test_{abs}$ |
| | No. 5 | 1 | Deployed | Roughly 50 | $train_{met}$/ $train_{abs}$ | $test_{abs}$ |
| | | 1 | Predeployed | 0 | $train_{met}$/ $train_{abs}$ | $test_{abs}$ |

no standardization of the echograph parameters was applied during the acquisitions. The pullback speed was 0.5 mm/s and the IVUS system had a 30 frames per second frame rate. The model of the metallic and bioabsorbable stents were "Promus Coronary Stent, Boston Scientific, MN" and "Absorb Bioresorbable Vascular Scaffold (BVS), Abbott, IL," respectively.

Two experts (one clinician and one experienced researcher in IVUS imaging) manually annotated the beginning and the end of the stent in each sequence. They were asked to scroll the pullback back and forward, looking at the short-axis view only, and analyze adjacent frames in order to identify the frames defining the stent boundaries. More than one annotation per pullback was allowed when several stents were implanted in subsequent segments of the same artery. Two separate test sets were defined ($test_{met}$ and $test_{abs}$), corresponding to pullback containing metalling or bioabsorbable stents, respectively. When sequences containing metallic stent were tested the training was performed using pullbacks having metallic stent ($train_{met}$). Instead when sequences containing bioabsorbable stent were analyzed, the training was performed separately using metallic ($train_{met}$) or bioabsorbable ($train_{abs}$) frames. Since the number of sequences containing a bioabsorbable stent was low, we expect that the stent detection performance might increase when the system is trained on a larger dataset. When pullbacks of the same dataset were used for test and training, a 10-fold cross-validation strategy was used to compute the results, while all the metallic pullbacks ($train_{met}$) were used for training purposes when the test were performed over $test_{abs}$ dataset.

### 3.3.2. Experiments on stent presence assessment

The assessment of stent presence is based on the analysis of the mono-dimensional signal $\gamma(t)$. In order to evaluate the performance, the manual annotations of beginning and end of the stent were converted into binary signals $\gamma_{obs}(t)$ indicating the presence of the stent in the pullback. Successively, the signals $\sqcap(t)$ indicating the segments of the pullback in which a stent is likely to be present were compared against the sections indicated by the observers $\gamma_{obs}(t)$. The performance was evaluated measuring the *Precision* ($P$), *Recall* ($R$), *F-measure* ($F$), and *Jaccard index* ($J$). The first three indices are typically used to evaluate a detection problem, while the forth index evaluate the overlap between the ground-truth and the automatically detected signals. Additionally, the distance between the automatic and the manual detection is expressed in number of gated frames, in mm, and in percentage of the stent length.

In our experiments, we estimate $\sigma_k^{train_{sax}}$, $\mu_k^{train_{sax}}$, and $L^{train_{sax}}$ by optimizing the F-measure score by varying $L_{sax}$ between 3 and 50. When the optimal value of $L_{sax} = 36$ is set, $\sigma_k^{train_{sax}}$, $\mu_k^{train_{sax}}$ are estimated automatically by the SAX algorithm.

**TABLE 5**
Quantitative Evaluation the Pullback Analysis Stage on Both $test_{met}$ and $test_{abs}$ Datasets.

| | | Precision Mean (std) | Recall Mean (std) | F-measure Mean (std) | Jaccard Mean (std) | Boundary err. [gated fr.] Mean (std) | Boundary err, relative Mean (std) |
|---|---|---|---|---|---|---|---|
| $test_{met}/$ $train_{met}$ | auto vsobs-1 | 91.0% (15.8%) | 87.4% (14.5%) | 87.5% (12.0%) | 79.6% (17.6%) | 4.0 (5.6) | 4.2% (3.8%) |
| | auto vsobs-2 | 94.3% (12.2%) | 79.3% (16.4%) | 84.6% (11.9%) | 75.0% (16.9%) | 5.3 (6.0) | 5.8% (4.4%) |
| | obs-1 vsobs-2 | 96.8% (14.1%) | 99.3% (5.4%) | 92.1% (10.5%) | 86.7% (14.5%) | 3.2 (5.0) | 3.9% (2.2%) |
| $test_{abs}/$ $train_{met}$ | auto vsobs-1 | 91.8% (12.8%) | 86.9% (13.4%) | 88.6% (11.1%) | 77.7% (13.7%) | 3.4 (4.8) | 10.4% (5.2%) |
| | auto vsobs-2 | 91.8% (12.5%) | 87.4% (12.7%) | 88.9% (10.4%) | 78.1% (12.8%) | 3.3 (4.9) | 10.3% (5.7%) |
| | obs-1 vsobs-2 | 99.5% (1.8%) | 99.0% (1.9%) | 99.2% (1.2%) | 98.5% (2.4%) | 2.9 (6.2) | 8.4% (7.4%) |
| $test_{abs}/$ $train_{abs}$ | auto vsobs-1 | 97.0% (9.0%) | 75.6% (22.9%) | 83.4% (16.5%) | 64.3% (15.6%) | 5.5 (7.5) | 17.8% (7.0%) |
| | auto vsobs-2 | 97.4% (7.7%) | 76.3% (22.9%) | 84.0% (16.2%) | 65.1% (15.9%) | 5.3 (7.2) | 16.9% (7.0%) |
| | obs-1 vsobs-2 | 99.5% (1.8%) | 99.0% (1.9%) | 99.2% (1.2%) | 98.5% (2.4%) | 2.9 (6.2) | 8.4% (7.4%) |

Notes: For each dataset the performance of the automatic method versus each manual annotation are reported. Then the interobserver variability is shown. For each experiment the Precision, Recall, F-measure, and Jaccard index are reported, along with the error in the boundary assessment, expressed in number of gated frames, and in percentage of the stent length.

The quantitative results for the pullback-wise analysis for the three experiments are reported in Table 5. In the first experiment, the system is trained and tested using metallic pullbacks. In the second and the third experiments, the test is performed over bioabsorbable frames, and the training is obtained using metallic and bioabsorbable data. As IVUS is highly challenging to interpret, the two observers sometimes disagree as reported in Table 5 (rows 5, 8, and 11).

In case of bioabsorbable stents, the best performances are obtained when the framework is trained using metallic frames. All the performance metrics (excepts the Precision) of the second experiment, that is, training using metallic frames (rows 6 and 7), are higher with respect to the results obtained in the third experiment, that is, training using bioabsorbable frames (rows 9 and 10). In particular, a 10% difference can be noted between the Recall and Jaccard measure of the two experiments. This confirms that the bioabsorbable dataset is too small for training purposes.

If we focus on the first two experiments (rows 3–8), in both datasets, the precision approaches the interobserver variability, while the recall is in general between 10% and 20% lower than the results of the manual annotation. The obtained F-measure and the Jaccard measure of the automatic performance show satisfactory results when compared with manual annotations (about 5% and 7% lower than the ground-truth score).

Such results are confirmed by analyzing the distance between the boundary assessed with our method and the manual annotations, as illustrated in the last two columns of Table 5. In the case of metallic stents, the error in the boundary assessment ranges between 4 and 5.3 gated frames (about 2 and 2.5 mm, respectively). It is interesting to note that such errors correspond on average to 4.2% and 5.8% of the stent length. In the case of metallic stents, the error in the boundary assessment is about 3.4 gated frames (about 1.5 mm). In this case, such error correspond in average to 10% of the stent length, given that the length of the bioabsorbable stents utilized in our dataset was in general smaller than the metallic ones. The results reported in Table 5 illustrate that the method is flexible and can be applied to pullback containing either metallic or bioabsorbable stents with similar results.

In order to analyze if the scores obtained by the automatic method were statistically different from the

**TABLE 6**

ANOVA Statistical Analysis of the Results Reported in Table 5, Indicating When the Performances of the Automatic Stent Detection Are Significantly Different With Respect to the Manual Annotations.

| | | Precision Mean (std) | Recall Mean (std) | F-measure Mean (std) | Jaccard Mean (std) | Boundary err. [gated fr.] Mean (std) | Boundary err, relative Mean (std) |
|---|---|---|---|---|---|---|---|
| $test_{met}$ $train_{met}$ | (auto vs obs−1) (obs−1 vs obs−2) | $P < .01$ | $P > .05$ | $.01 < P < .05$ | $.01 < P < .05$ | $P > .05$ | $P > .05$ |
| | (auto vs obs−2) (obs−1 vs obs−2) | $P < .01$ | $P < .01$ | $P < .01$ | $P < .01$ | $P < .01$ | $P > .05$ |
| $test_{abs}$ $train_{met}$ | (auto vs obs−1) (obs−1 vs obs−2) | $P < .01$ | $.01 < P < .05$ | $P < .01$ | $P < .01$ | $P < .01$ | $P > .05$ |
| | (auto vs obs−2) (obs−1 vs obs−2) | $P < .01$ | $.01 < P < .05$ | $P < .01$ | $P < .01$ | $P < .01$ | $P > .05$ |
| $test_{abs}$ $train_{abs}$ | (auto vs obs−1) (obs−1 vs obs−2) | $P < .01$ | $P < .01$ | $P < .01$ | $P < .01$ | $P < .01$ | $P < .01$ |
| | (auto vs obs−2) (obs−1 vs obs−2) | $P < .01$ | $P < .01$ | $P < .01$ | $P < .01$ | $P < .01$ | $P < .01$ |

Notes: A *strong* or *weak statistical difference* between the results is considered when the $P$-value is <.01 or between $.01 < P < .05$, respectively. Otherwise, when the $P$-value is >.05, the null hypothesis of statistical difference between the results cannot be rejected, hence indicating that the two performances can be considered *comparable*. The second column indicated which performances of Table 5 are considered in the ANOVA analysis.

performance of the manual annotation, an ANOVA test was performed over the results reported in Table 5. As can be observed in Table 6, the performances of the method were particularly satisfying, since in the case of metallic stent the automatic method approaches the performances of one observer. The error of boundary detection committed by the automatic method was found comparable with respect the errors committed by the first observer, and the difference of F-measure and Jaccard scores committed by the automatic method with respect to the manual annotation was statistically weak ($.01 < P < .05$). On the other hand, the results were found statistically different ($P < .01$) when the automatic method was compared against the manual annotation of the second observer.

In the case of bioaborbable frames trained using metallic pullbacks, comparable results were found only in the relative boundary errors, while in the other cases the results between automatic and manual detection was considered statistically different. Finally, in the case of bioaborbable frames trained using bioaborbable pullbacks, all the test reported a statistical difference between the automatic and the manual detection.

### 3.3.3. Malapposition analysis

Malapposition happens when at least one stent strut is separated from the intimal surface of the arterial wall and is generally computed as the distance between the malapposed struts and the vessel wall, and the area is measured.

In particular, the thresholds of malapposed struts depend on the stent type and brand[38] ranging from 100 to 160 μm. In the case of the Boston Scientific brand, we consider the stent malapposed when the distance to the lumen is higher than 130 μm. The framewise analysis obtained applying the framework to each IVUS gated frame allows to estimate the lumen and stent contours. Using such information it is possible to compute the stent malapposition for each angular sector of the frame along the pullback. As shown in Fig. 11 (top), an exemplar IVUS pullback is analyzed. The framework provides a map indicating the distance between the lumen and the stent in each angular section of the pullback. Then a plot in Fig. 11 (middle) illustrates the maximum amplitude of the distance along the pullback. Finally, a third plot indicates the percentage of the stent displaying malapposition (Fig. 11 (bottom)).

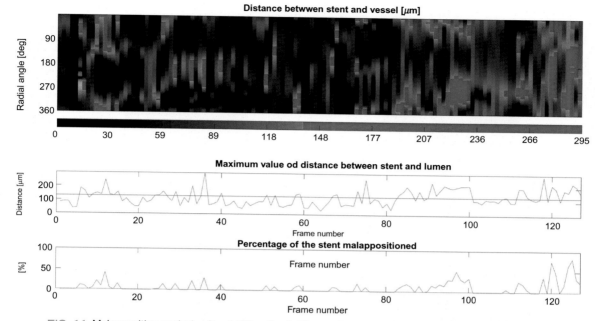

**FIG. 11** Malapposition analysis of an IVUS pullback. The *top image* represents a distance map between the stent and the lumen. The *vertical axis* represents the angular section of the IVUS image, while the *horizontal axis* displays the position of the frame. The intensity of the map corresponds to the distance in μm. When the distance is superior to the malapposition threshold a *red color* is displayed. The *second plot* displays the maximum amplitude of the distance between the stent and the lumen, along the frames of the pullback; a *horizontal line* indicates the malapposition threshold. The *third plot* displays the percentage of the stent malapposed in each frame of the pullback.

## 3.4. Results
### 3.4.1. Metallic stents
Examples of processed signals for the detection of metallic stent in the IVUS frame are depicted in Fig. 12. In Fig. 12A1 and B1, both initial and final frame of the sequence are accurately identified. The result is not obvious since in (a) the amplitude of the signal $\gamma(t)$ is almost null in two sections of the pullback. However, the SAX algorithm allowed to detect the presence of stent, based on the statistics of the frames in the neighborhood. On the other hand, in Fig. 12A2, the central section of the pullback where $\gamma_{auto}(t)$ is almost null is correctly classified by the SAX algorithm as the absence of stent. This is coherent with the manual annotation of the two observers, where two stents are labeled. It must be noted that in this chapter no constrain on the minimum length of the stent is applied. In Fig. 12B2, we observed a case in which the absence of the stent is correctly identified. Indeed, observing Fig. 12B2, it can be noted that the quantification of the signal is below the global threshold $\mu_{N_{sax}}^{train_{pull}}$. In Fig. 12A3 and B3, regions of high signal separated from the main stent have been identified as

a secondary implanted stent. It might be noted that this error happens only when a strong spike in the signal is present, for instance, when a calcified plaque is mistaken for a deployed stent. In Fig. 12A4 and B4, the initial and the final frame of the stent are incorrectly identified since the amplitude of $\gamma(t)$ is low and comparable with the noise at the proximity of the stent area.

**Bioabsorbable stents.** Examples of processed signals for the detection of bioadsorbable stent in the IVUS frame are depicted in Fig. 13. In Fig. 13A1 and B1, we observe two cases of accurate stent detection. In the rest of cases, the performance of the stent detection decreases. In Fig. 13A2, B2, and A3, the main stent was accurately detected, however, one or more small peaks in the signal, having high magnitude, are incorrectly classified. In Fig. 13A3, it is interesting to note how the algorithm is able to recover the whole stent longitude at the right side of the image, even if the signal amplitude is low at the middle of the stent area. Finally, Fig. 13B3, A4, and B4 show pullback not containing a

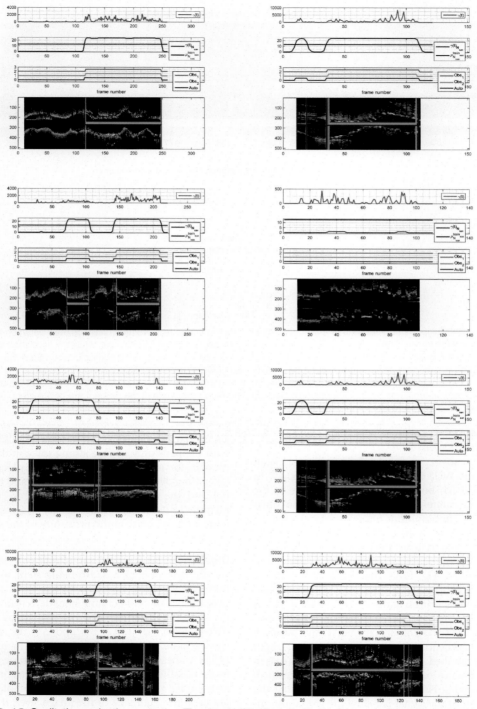

**FIG. 12** Qualitative evaluation on $test_{met}$. The signal $\gamma(t)$ is illustrated in the first row, while in the second row, the result of the SAX quantization is reported. Finally, in the third row, three binary signals representing the presence or the absence of the signal are compared: the first correspond to the automatic results, while the second and the third are the annotations of the two observers.

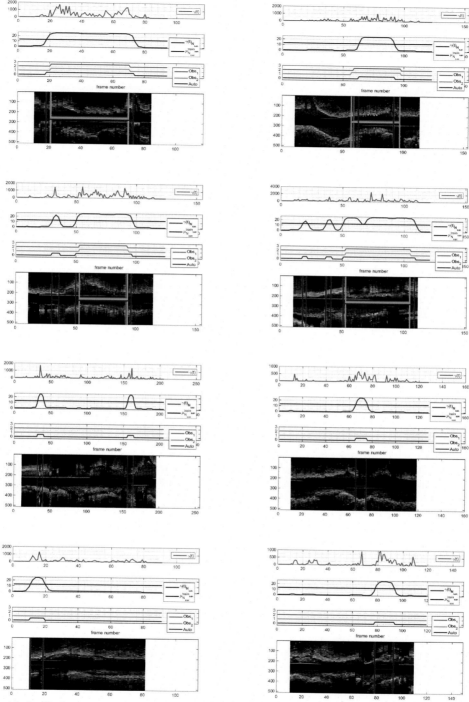

FIG. 13 Qualitative evaluation on $test_{abs}$ dataset. The signal $\gamma(t)$ is illustrated in the first row, while in the second row, the result of the SAX quantization is reported. Finally, in the third row, three binary signals representing the presence or the absence of the signal are compared: the first corresponds to the automatic results, while the second and the third are the annotations of the two observers.

stent in which short but intense peaks in the signal are misclassified.

## 3.5. Discussion

In this chapter, a framework for the automatic identification of stent presence along the pullback (location and extension) has been presented.

The methodology described in the chapter is based on three main steps: (1) a gating, (2) a stent-strut detection techniques used to obtain a measure of likelihood for a frame to contain a stent, and (3) a new strategy, introduced in this chapter, for obtaining a robust binary representation of the presence of the stent in the pullback using the SAX algorithm. Stages (1) and (2) of the pipeline can be computed using a similar method proposed in the state of the art.[9],[15–19] However, since the reliability of the strut detection is a critical step of the pipeline, the influence of using a method less reliable might have a negative impact over the performances of the whole framework.

In clinical practice, the presented system could be used intraoperatively to perform automatic analysis of the stent position and placement. This would ease the task of physicians, by reducing the burden of manual search and measurements of the deployed stent via inspection of the IVUS sequence. In order to show how the result can be represented in a clinical application, in Fig. 14 two exemplar IVUS long-axis view of the stent detection are depicted, superimposed with vertical lines representing the boundaries of the stent manually labeled by the physician.

The presented framework process a sequence of gated frames. Due to the catheter swinging effect, nongated frames might be analyzed multiple times along the sequence. Application of gating has the advantage of providing frames that are less affected by motion artifacts, in which the analysis of blood texture can be done in a more robust and reproducible way. Moreover, gating reduces the amount of frames for analysis by approximately a factor 30, which also benefits the system in terms of computation time. The use of gated frames does not limit the applicability of the proposed system, which can potentially work on nongated frames as well. However, in this case a reduction in performance can be expected.

The analysis of the *stent presence* signal has been performed using the SAX algorithm, which provides an unsupervised classification of the stent location in a fast and statistically robust manner.

Since the proposed system is meant to be used in an intraoperative manner, it is important to evaluate the computational cost in order to assess its real impact during clinical practice. For this purpose, we evaluated the computation time of the system. The method has been implemented in Matlab (The MathWorks) and the computation time of the pullback analysis is one order of magnitude lower than the time required for detecting the stents.[9] Indeed the pullback-wise analysis took about 4.1 s per pullback versus 0.33 s per frame (around 33 s per pullback) required for a stent detection (measured on a Intel i7 quad-core processor).

## 3.6. Conclusion

We presented a new strategy for the assessment of stent location and extension of intracoronary stent in IVUS sequences. Such technique, when coupled with previous published methods,[9],[22] provides a complete framework for the intracoronary stents aimed at assisting intraoperative diagnosis. The results obtained on an heterogeneous dataset containing both metallic and bioabsorbable stents collected through a multicentric collaboration are close to the interobserver variability and suggest that the system has the potential to be used during percutaneous interventions.

In the case of metallic stent, the performances are an overall F-measure of 87.4% and 86.4% and a Jaccard score 76.6% and 75.0% obtained comparing the automatic results against the manually annotated struts of the two observers, respectively. For bioabsorbable stents, an overall F-measure of 88.6% and 88.9% were obtained, with a Jaccard score of 77.7% and 78.1% when compared with the automatic results against the manually annotated struts. Such values approached the interobserver variability which corresponded to an F-measure of 92.1% and a Jaccard score of 86.7% in the case of metallic stents, and an F-measure of 99.2% and a Jaccard score of 98.5% in the case of bioabsorbable stents.

With regard to the distance between the automatic and the manual detection, the error in case of metallic stents is 4.0 and 5.3 gated frames (corresponding to about 2 and 2.5 mm, respectively) from the manually annotated struts of the two observers, while in case of bioabsorbable stents the error is 3.4 and 3.3 gated frames (corresponding to about 1.5 mm). These results are satisfying considering that the interobserver error of the manual annotation is 3.2 and 2.9 gated frames, in case of metallic and bioaborbable stents, respectively.

The system has been validated using cases of both metallic and bioabsorbable stents. The performances of the system were evaluated when metallic or bioabsorbable stents were used for training, and the best performances were obtained using data from cases with metallic stents, since dataset was larger. Bioabsorbable

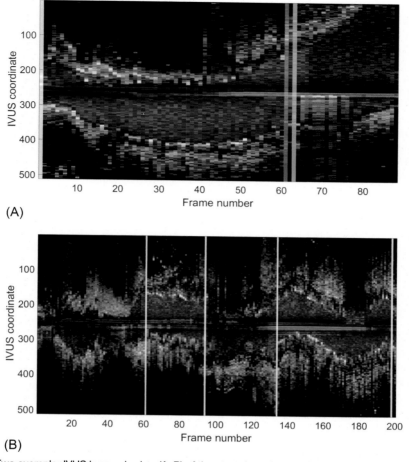

**FIG. 14** Two exemplar IVUS long-axis view (A, B) of the stent detection are depicted, superimposed with *red and yellow vertical lines* representing the boundaries of the stent manually labeled by the experts. The *stent presence* ⊓ (*t*) automatically computed using the SAX algorithm was indicated in *blue* and, for each frame, only the stent area computed by the automatic method[9] was colored. The pictures illustrate how the results can be represented in a clinical application.

stents are used less often during PCIs, therefore finding a large set of examples was hard. However, good performance can be observed also in the presence of bioabsorbable stents, showing the capability of the system to generalize well to detection of unseen examples, including other types of stent.

Future work will be addressed toward comparing the performance of the current pipeline against the assessment of stent location and extension obtained using deep learning (DL) techniques. In particular, artificial neural networks (ANN) might improve the classification on both frame-wise and pullback-wise analyses. ANNs are well known for providing state-of-the-art classification performances; however, it should be noted that ANN are currently affected by some limitations: (1) a large dataset (of the order of 10,000 samples) is usually required for training purposes. (2) In literature, there is no clear consensus about how to find the optimal parameter for the network fine-tuning. Such limitations should be carefully studied in order to analyze if a deep learning strategy could be successfully applied to IVUS pullbacks.

## 4. GENERAL CONCLUSION

In this chapter, a fully automatic framework for CAD of intracoronary stents in IVUS image sequences is presented. This CAD system combines two applications: the first technique aimed at *stent shape estimation*, and the second method aimed at the *longitudinal stent location*, which exploit the results of the shape estimation to detect the boundaries and the position of the stent along the IVUS pullback. In both applications, the results obtained comparing the automatic results with the manual annotation of two observers show that the method approaches the interobserver variability. Similar performances were obtained for both metallic and bioabsorbable stents, showing the flexibility and robustness of the method. As a general conclusion, the CAD system described in this chapter might be considered as promising tool for the stent analysis in IVUS sequences, and might be used in clinical practice in the future.

## ACKNOWLEDGMENTS

The authors would like to thank Dr. Eleazar García Díaz from the CCR Ascardio Center, Barquesimeto (Venezuela), Dr. Joan Antoni Gomez, and Dr. Angel Cequier from the Bellvitge University Hospital, Barcelona (Spain), Dr. Alfonso Medina from University Hospital Dr. Negrin, Gran Canaria (Spain), and Dr. Hany Ragy from National Heart Institute, Imbaba (Egypt) for providing part of the data used in this study. This study was supported in part by the Grant RTI2018-095232-B-C21, 2017 SGR 1742, CERCA, and ICREA Academia 2014. The authors have no relevant conflicts of interest to disclose. The authors gratefully acknowledge the support of NVIDIA Corporation with the donation of the Titan Xp used in our research.

## REFERENCES

1. Trabattoni D, Bartorelli AL. IVUS in bifurcation stenting: what have we learned? *EuroIntervention*. 2010; 6(suppl J): J88–J93. https://doi.org/10.4244/EIJV6SUPJA14.
2. Yoon HJ, Hur SH. Optimization of stent deployment by intravascular ultrasound. *Korean J Intern Med*. 2012; 27 (1):30–38. https://doi.org/10.3904/kjim.2012.27.1.30.
3. de Lezo JS, Medina A, Martín P, Novoa J, Pan M, Caballero E. Predictors of ostial side branch damage during provisional stenting of coronary bifurcation lesions not involving the side branch origin: an ultrasonographic study. *EuroIntervention*. 2012; 7(10):1147–1154.
4. Andell P, Karlsson S, Mohammad MA, et al. Intravascular ultrasound guidance is associated with better outcome in patients undergoing unprotected left main coronary artery stenting compared with angiography guidance alone. *Circ Cardiovasc Interv*. 2017; 10(5):e004813.
5. Alberti M, Balocco S, Gatta C, et al. Automatic bifurcation detection in coronary IVUS sequences. *IEEE Trans Biomed Eng*. 2012; 59(4):1022–2031.
6. Waggoner J, et al. How do OCT and IVUS differ? A comparison and assessment of these modern imaging modalities. *Card Interv Today*. 2011; 3:46–52.
7. Reiber JHC, Tu S, Tuinenburg JC, Koning G, Janssen JP, Dijkstra J. QCA, IVUS and OCT in interventional cardiology in 2011. *Cardiovasc Diagn Therapy*. 2011; 1 (1):57–70.
8. Waksman R, Kitabata H, Prati F, Albertucci M, Mintz GS. Intravascular ultrasound versus optical coherence tomography guidance. *J Am Coll Cardiol*. 2013; 62 (17 suppl):S32–S40. https://doi.org/10.1016/j.jacc.2013. 08.709.
9. Ciompi F, Balocco S, Rigla J, Carrillo X, Mauri J, Radeva P. Computer-aided detection of intracoronary stent in intravascular ultrasound sequences. *Med Phys*. 2016; 43 (10):5616–5625.
10. Onuma Y, Serruys PW. Bioresorbable scaffold: the advent of a new era in percutaneous coronary and peripheral revascularization? *Circulation*. 2011; 123(7):779–797. https://doi.org/10.1161/ CIRCULATIONAHA.110.971606.
11. Gogas BD, Farooq V, Onuma Y, Serruys PW. The ABSORB bioresorbable vascular scaffold: an evolution or revolution in interventional cardiology? *Hellenic J Cardiol*. 2012; 53 (4):301–309.
12. Wang Z, Jenkins MW, Linderman GC, et al. 3-D stent detection in intravascular OCT using a Bayesian network and graph search. *IEEE Trans Med Imaging*. 2015; 34 (7):1549–1561.
13. Lu H, Gargesha M, Wang Z, et al. Automatic stent strut detection in intravascular OCT images using image processing and classification technique. *SPIE Medical Imaging*. International Society for Optics and Photonics; 2013:867015.
14. Wang A, Eggermont J, Dekker N, et al. Automatic stent strut detection in intravascular optical coherence tomographic pullback runs. *Int J Cardiovasc Imaging*. 2013; 29(1):29–38.
15. Dijkstra J, Koning G, Tuinenburg JC, Oemrawsingh PV, Reiber JHC. Automatic stent border detection in intravascular ultrasound images. *International Congress Series*; vol. 1256:Elsevier; 2003:1111–1116.
16. Canero C, Pujol O, Radeva P, et al. Optimal stent implantation: three-dimensional evaluation of the mutual position of stent and vessel via intracoronary echocardiography. *Computers in Cardiology*. 1999: 261–264. https://doi.org/10.1109/CIC.1999.825956.
17. Dijkstra J, Koning G, Tuinenburg JC, Oemrawsingh PV, Reiber JHC. Automatic border detection in intravascular ultrasound images for quantitative measurements of the vessel, lumen and stent parameters. *Computers in Cardiology (Cat No 01CH37287)*. 28:IEEE; 2001:25–28 1230.

18. Rotger D, Radeva P, Bruining N. Automatic detection of bioabsorbable coronary stents in IVUS images using a cascade of classifiers. *IEEE Trans Inf Technol Biomed.* 2010; 14(2):535–537. https://doi.org/10.1109/TITB.2009. 2017528.

19. Hua R, Pujol O, Ciompi F, et al. Stent strut detection by classifying a wide set of IVUS features. *MICCAI Workshop on Computer Assisted Stenting.* 2012:130–137.

20. Ciompi F, Pujol O, Gatta C, et al. Holimab: a holistic approach for media-adventitia border detection in intravascular ultrasound. *Med Image Anal.* 2012; 16:1085–1100.

21. Gatta C, Puertas E, Pujol O. Multi-scale stacked sequential learning. *Pattern Recogn.* 2011; 44(10-11):2414–2426.

22. Gatta C, Balocco S, Ciompi F, Hemetsberger R, Rodriguez-Leor O, Radeva P. Real-time gating of IVUS sequences based on motion blur analysis: method and quantitative validation. *MICCAI 2010, LNCS 6362/2010.* 2010:59–67.

23. Balocco S, Gatta C, Ciompi F, et al. Combining growcut and temporal correlation for IVUS lumen segmentation. *IbPRIA, LNCS 6669/2011.* 2011:556–563.

24. Ciompi F, Hua R, Balocco S, et al. Learning to detect stent struts in intravascular ultrasound. *Pattern Recognition and Image Analysis.* Springer; 2013:575–583.

25. Ciompi F, Balocco S, Caus C, Mauri J, Radeva P. Stent shape estimation through a comprehensive interpretation of intravascular ultrasound images. *Medical Image Computing and Computer-Assisted Intervention-MICCAI 2013.* Springer; 2013:345–352.

26. Schapire R. The boosting approach to machine learning: an overview. *MSRI Workshop on Nonlinear Estimation and Classification, Berkeley, CA, USA.* 2001.

27. Dietterich TG, Bakiri G. Solving multiclass learning problems via error-correcting output codes. *JAIR.* 1995; 2:263–286.

28. Balocco S, Gatta C, Ciompi F, et al. Standardized evaluation methodology and reference database for evaluating IVUS image segmentation. *Comput Med Imaging Graph.* 2014; 38(2):70–90. https://doi.org/10.1016/j.compmedimag.2013.07.001.

29. Gatta C, Pujol O, Rodriguez-Leor O, Mauri J, Radeva P. Fast rigid registration of vascular structures in IVUS sequences. *IEEE Trans Inf Technol Biomed.* 2009; 13(6):1006–1011.

30. Tsantis S, Kagadis GC, Katsanos K, Karnabatidis D, Bourantas G, Nikiforidis GC. Automatic vessel lumen segmentation and stent strut detection in intravascular optical coherence tomography. *Med Phys.* 2012; 39 (1):503–513. https://doi.org/10.1118/1.3673067.

31. Tenekecioglu E, Albuquerque FN, Sotomi Y, et al. Intracoronary optical coherence tomography: clinical and research applications and intravascular imaging software overview. *Catheter Cardiovasc Interv.* 2017; 89(4):679–689.

32. Mintz GS, Pichard AD, Satler LF, Popma JJ, Kent KM, Leon MB. Three-dimensional intravascular ultrasonography: reconstruction of endovascular stents in vitro and in vivo. *J Clin Ultrasound.* 1993; 21 (9):609–615.

33. von Birgelen C, Mintz GS, Nicosia A, et al. Electrocardiogram-gated intravascular ultrasound image acquisition after coronary stent deployment facilitates on-line three-dimensional reconstruction and automated lumen quantification. *J Am Coll Cardiol.* 1997; 30 (2):436–443.

34. Umemoto T, Pacchioni A, Nikas D, Reimers B. Recent developments of imaging modalities of carotid artery stenting. *J Cardiovasc Surg.* 2017; 58(1):25–34.

35. Kaple RK, Tsujita K, Maehara A, Mintz GS. Accuracy of stent measurements using ECG-gated greyscale intravascular ultrasound images: a validation study. *Ultrasound Med Biol.* 2009; 35(8):1265–1270.

36. Balocco S, Ciompi F, Rigla J, Carrillo X, Mauri J, Radeva P. Assessment of intracoronary stent location and extension in intravascular ultrasound sequences. *Med Phys.* 2019; 46(2):484–493.

37. Lin J, Keogh E, Wei L, Lonardi S. Experiencing SAX: a novel symbolic representation of time series. *Data Min Knowl Disc.* 2007; 15(2):107. https://doi.org/10.1007/s10618-007-0064-z.

38. Lee SY, Hong MK. Stent evaluation with optical coherence tomography. *Yonsei Med J.* 2013; 54(5):1075–1083.

# CHAPTER 11

# Real-Time Robust Simultaneous Catheter and Environment Modeling for Endovascular Navigation <sup>☆</sup>

LIANG ZHAO[a] • STAMATIA GIANNAROU[b] • SU-LIN LEE[c] • GUANG-ZHONG YANG[d]
[a]Center for Autonomous Systems, University of Technology Sydney, Ultimo, NSW, Australia,
[b]Hamlyn Center for Robotic Surgery, Imperial College London, London, United Kingdom,
[c]EPSRC Center for Interventional and Surgical Sciences, University College London, London, United Kingdom, [d]Institute of Medical Robotics, Shanghai Jiao Tong University, Shanghai, China

## 1. INTRODUCTION

Cardiovascular diseases (CVD) are the number one cause of death globally, taking an estimated 17.9 million lives each year. Endovascular catheter procedures are among the most common surgical interventions used for minimally invasive intervention of cardiovascular diseases.[1–5] Despite their benefit in providing minimal trauma to the patient, manipulating catheters through the fragile yet complex endovascular system in the presence of physiological motion is a challenging task.[6–8] Endovascular navigation still rely on 2D guidance based on X-ray fluoroscopy together with the use of contrast agents.[9,10] To establish safer endovascular procedures, knowledge of the interaction between the catheter and its surroundings is required and thus the 3D structure of the vasculature needs to be recovered intraoperatively. The current clinical approach is to overlay 3D vessel models reconstructed from preoperative imaging but due to the dynamic nature of the vasculature, these models must be updated in real time, intraoperatively.

Recently, intravascular ultrasound (IVUS) has been used as an imaging guidance to reduce the radiation of fluoroscopy and use of nephrotoxic contrast agent. However, the challenge associated with vessel reconstruction is that the cross-sectional data of IVUS will change significantly according to the motion of the imaging tip because of the limited field of view of IVUS. Initial approaches reconstruct the vessel by supposing that IVUS images are parallel to each other.[11,12] This assumption can only be applied to the IVUS pullback data, in which the image sequence is recorded as the catheter is slowly withdrawn from the vessel. In practice, this is not realistic in areas where the curvature of the vessel is relatively large. The registration of IVUS images to angiography data for vessel reconstruction has been explored to align the images in 3D space[13–16] but this beats the purpose of minimizing X-ray radiation and the use of contrast agent. In the meanwhile, catheter pose estimated in these approaches is sensitive to noises, thus additional sensing information is required to accurately reconstruct the vessel. To this end, IVUS and electromagnetic (EM) sensing data have been fused for intraoperative vessel reconstruction.[17,18] However, existing methods treated observations from IVUS and EM as exact and their uncertainty was ignored. This makes the method vulnerable to errors in the observations which can affect the accuracy of the recovered vessel structure. For instance, vessel boundaries do not always appear complete on IVUS images and EM sensors are prone to measurement errors.

In this chapter, a robust real-time 3D vessel reconstruction algorithm for endovascular catheter control and navigation is proposed based on IVUS and EM sensing. The 3D vessel reconstruction is formulated as a nonlinear optimization problem by considering the uncertainty in both the IVUS contour and the EM pose, as well as vessel morphology provided by preoperative data. To enable vessel reconstruction in real time, an approximation to the optimization problem is proposed in which the objective function and Jacobian related to solving the nonlinear optimization can be

☆This work was supported by the Commissions 7th Framework Program FP7-ICT Project "Cognitive AutonomouS CAtheter operating in Dynamic Environments (CASCADE)," under grant agreement no. 601021.

*Intravascular Ultrasound.* https://doi.org/10.1016/B978-0-12-818833-0.00011-4

computed a priori on the preoperative data. Detailed validation on phantom datasets is performed to demonstrate the accuracy of the proposed algorithm and its robustness to abrupt catheter motions.

## 2. VESSEL MODELING IN SCEM

Simultaneous catheter and environment modeling (SCEM), as described in Ref. 17, proposes a reconstruction of the vasculature based on data fusion from intravascular ultrasound (IVUS) and EM tracking. In this framework, anatomical information is extracted by segmenting the contour of the inner vessel wall $C_I$ from the IVUS images that consists of a set of boundary points and is defined as

$$C_I = [c_1^{I^T}, ...., c_n^{I^T}]^T,  \qquad (1)$$

where each boundary point can be presented as

$$c_j^I = [x_j^I, y_j^I, 0]^T, \quad j = 1, ..., n. \qquad (2)$$

Here, the boundary point extracted from the IVUS image $[x_j^I, y_j^I]^T$ is padded with a 0 as the z coordinate to make $c_j$ a 3D point in the IVUS coordinate frame. The pose $P_E$ captured by the EM sensor attached to the tip of the IVUS catheter is used to transform the IVUS contour $C_I$ to a global coordinate system by

$$c_j = R_E^T c_j^I + T_E, \qquad (3)$$

where $R_E$ is the $3 \times 3$ rotation matrix and $T_E$ is the $3 \times 1$ translation vector of the EM pose $P_E$. $C = [c_1^T, ...., c_n^T]^T$ is the IVUS contour transformed into the global coordinate frame. Thus, the reconstruction of the vessel is obtained by transforming and combining all the IVUS contours extracted during vessel scanning to a common global coordinate system.

However, in SCEM, both the measurements from IVUS and EM are considered as exact, and any error associated with the measurements will transfer into the vessel reconstruction (see Fig. 1). The EM poses and IVUS contours are always associated with significant measurement errors which should not be ignored for accurate vessel reconstruction. For instance, the IVUS data is noisy and the vessel boundaries do not always appear complete on IVUS images, making the contour extraction challenging.

## 3. METHODS

The proposed vessel reconstruction framework and the improvements to the reconstruction is shown in Fig. 2. In this chapter, both the contours from IVUS and poses from EM are considered as measurements with errors and uncertainties and combined with the vessel

structure provided by preoperative data to formulate a nonlinear optimization problem.

### 3.1. Optimization Based on the Preoperative Data

The proposed nonlinear optimization estimates the optimized catheter pose $\hat{P}$ that will recover an accurate vessel structure. In the optimization, the IVUS contour $C_I$ and EM pose $P_E$ are considered as observations with uncertainties defined by the covariance matrices $\Sigma_I$ and $\Sigma_E$, respectively, and the catheter pose $P = [R, T]$ is defined as the state vector. The nonlinear least squares problem can then be mathematically formulated as

$$\underset{P}{\mathrm{argmin}} \sum_{j=1}^n \| c_j^I - f_j(P) \|_{\Sigma_j^{-1}}^2 + \| P_E - P \|_{\Sigma_E^{-1}}^2, \qquad (4)$$

where

$$f_j(P) = R(c_j^M - T). \qquad (5)$$

The first term in Eq. (4) aims to minimize the difference between the contour $C_I$ from the IVUS image and the contour $f(P) = [f_1(P)^T, ...., f_n(P)^T]^T$ calculated from the preoperative data, weighted by the uncertainty of the IVUS contour $\Sigma_I = \mathrm{diag}(\Sigma_1, ..., \Sigma_n)$. The contour $C_M = [c_1^M T, ...., c_n^M T]^T$ is extracted from the preoperative data as the cross-section of the CT model and the plane defined by the catheter pose $P$. The function $f(\cdot)$ transforms $C_M$ from the global into the IVUS coordinate system. The second term in Eq. (4) minimizes the difference between the catheter pose $P$ and the pose $P_E$ obtained from EM, weighted by the uncertainty of the EM pose $\Sigma_E$.

A solution $\hat{P}$ of Eq. (4) can be obtained with the Gauss-Newton method,[19] starting with an initial estimate $P_0$ and iterating with $P_{k+1} = P_k + \Delta_k$. The vector $\Delta_k$ is the solution to

$$\left( \sum_{j=1}^n J_j^T \Sigma_j^{-1} J_j + \Sigma_E^{-1} \right) \Delta_k = \sum_{j=1}^n J_j^T \Sigma_j^{-1} (c_j^I - f_j(P)) + \Sigma_E^{-1}(P_E - P),$$

$$(6)$$

where $J_j$ is the linear mapping represented by the Jacobian matrix $\partial f_j / \partial P$ evaluated at $P_k$. Given that $C_M(P)$ is not constant but a function of $P$, the Jacobian $J_j$ can be computed as

$$J_j = \frac{\partial f_{c_j^M}}{\partial P} + R \frac{\partial c_j^M}{\partial P} \qquad (7)$$

where $\partial f_{c_j^M} / \partial P$ is the Jacobian with respect to $P$ when considering $C_T$ as constant. The Jacobian of $C_M$ w.r.t pose $P$ can be computed numerically on the preoperative data by using the finite difference method[20]

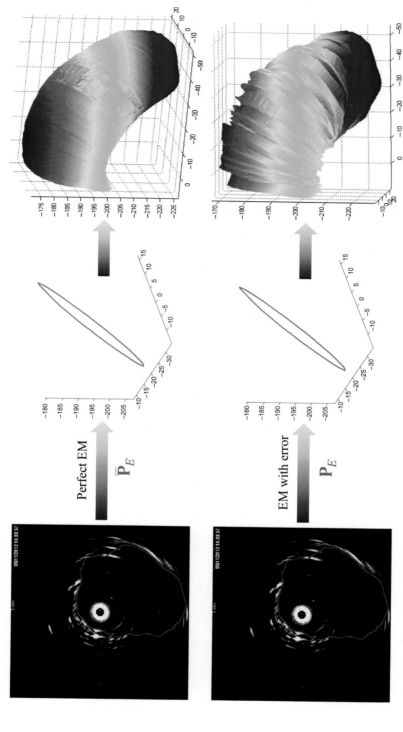

FIG. 1 The framework proposed in SCEM.[17] The errors from IVUS/EM will be directly transferred into the 3D vessel reconstruction.

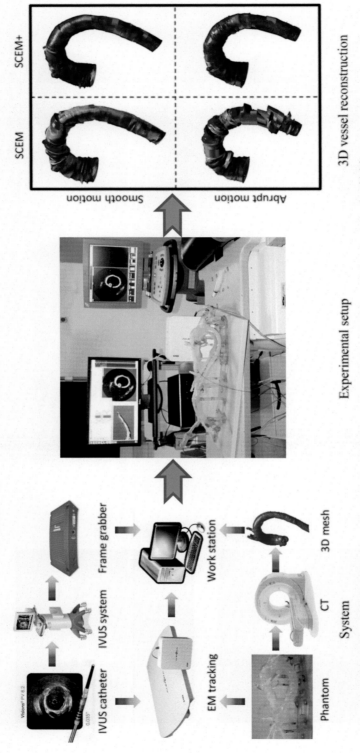

**FIG. 2** The proposed 3D vessel reconstruction setup (*left*) and the proposed improvements to the reconstruction (*right*).

$$\frac{\partial c_j^M}{\partial \mathbf{P}} = \frac{c_j^M(\mathbf{P}+\Delta) - c_j^M(\mathbf{P}-\Delta)}{2\Delta}, \tag{8}$$

where $\Delta$ is a small change on the state vector $\mathbf{P}$.

After an optimal solution of Eq. (4) is obtained, the uncertainty of the estimated catheter pose $\hat{\mathbf{P}}$ is given by the covariance matrix

$$\Sigma_P^{-1} = \sum_{j=1}^{n} J_j^T \Sigma_j^{-1} J_j + \Sigma_E^{-1}. \tag{9}$$

Then, the 3D vessel shape can be recovered by transforming $C_I$ to the global coordinate frame $C = [c_1^T, ..., c_n^T]^T$ using the optimized $\hat{\mathbf{P}} = [\hat{R}, \hat{\mathbf{T}}]$, with reconstruction uncertainty $\Sigma_C$:

$$c_j = \hat{R}^T c_j^I + \hat{\mathbf{T}}, \quad \Sigma_C = J_C \Sigma J_C^T, \tag{10}$$

where

$$J_C = \begin{bmatrix} \frac{\partial C}{\partial \mathbf{P}} & \frac{\partial C}{\partial C_I} \end{bmatrix} \tag{11}$$

and

$$\frac{\partial C}{\partial \mathbf{P}} = \begin{bmatrix} \frac{\partial c_1}{\partial \mathbf{P}} \\ \vdots \\ \frac{\partial c_n}{\partial \mathbf{P}} \end{bmatrix}, \quad \frac{\partial C}{\partial C_I} = \begin{bmatrix} \frac{\partial c_1}{\partial c_1^I} & \cdots & 0 \\ \vdots & \ddots & \vdots \\ 0 & \cdots & \frac{\partial c_n}{\partial c_n^I} \end{bmatrix} \tag{12}$$

is the Jacobian of $C$ w.r.t the catheter pose $\mathbf{P}$ and IVUS contour $C_I$, respectively, and

$$\Sigma = \begin{bmatrix} \Sigma_P & \\ & \Sigma_I \end{bmatrix}. \tag{13}$$

## 3.2. Uncertainty Estimation

To deal with the errors in the IVUS and EM measurements, the terms in the optimizations in Eq. (4) are weighted with the uncertainties $\Sigma_I$ and $\Sigma_E$. The uncertainty of the EM pose $\Sigma_E$ can be directly obtained from the position and orientation accuracy of the EM system. To estimate the uncertainty of the IVUS contour $\Sigma_I$, the contour is represented as a set of boundary points $C_I = \{c_j^I\}, j=1,...,n$ and the covariance matrix of $C_I$ can be computed as

$$\Sigma_I = \begin{bmatrix} \Sigma_1 & \cdots & 0 \\ \vdots & \ddots & \vdots \\ 0 & \cdots & \Sigma_m \end{bmatrix}, \quad \text{where } \Sigma_j = \delta_j^{-1} I. \tag{14}$$

Here, the points of the contour are assumed to be independent, hence $\Sigma_I$ is block diagonal. The covariance matrix of each point $\Sigma_j$ is computed by the weight of

the point $\delta_j$. The covariance matrix computed in this way is spherical.

In this chapter, the intensities of the points around the extracted contour on the IVUS image are used to compute the weights $\delta_j$. The contour points $c_j^I$ are first transformed from the Cartesian coordinates to polar coordinates $(\rho_j, \phi_j)$. Afterwards, the intensities of all the image points on the direction $\phi_j$ within a distance $m$ from $\rho_j$ in the polar coordinate system are accumulated. And the weight $\delta_j$ can be computed as

$$\delta_j = \begin{cases} 1, & \text{if } \gamma_j \geq t \\ \gamma_j/t, & \text{if } \gamma_j < t, \end{cases} \tag{15}$$

where

$$\gamma_j = \sum_{l=0}^{m} V((\rho_j+l)\cos\phi_j, (\rho_j+l)\sin\phi_j). \tag{16}$$

Here $V$ is the IVUS image. According to Eq. (15), if $\gamma_j$ is greater than a threshold $t$, the point $F_j$ is considered to lie on the vessel wall and its weight is set to 1. We assume that if a point has no high-intensity neighbors, it does not belong to the vessel wall and its weight is set to 0. An example of the uncertainty of IVUS contour is illustrated in Fig. 3.

## 3.3. Real-Time Implementation

To enable vessel reconstruction in real time, the original objective function of the proposed optimization scheme can be modified by multiplying $R^T$ on the first term:

$$\arg\min_{\mathbf{P}} \sum_{j=1}^{n} \| c_j^M - g_j(\mathbf{P}) \|_{\Sigma_j^{-1}}^2 + \| \mathbf{P}_E - \mathbf{P} \|_{\Sigma_E^{-1}}^2, \tag{17}$$

where

$$g_j(\mathbf{P}) = R^T c_j^I + \mathbf{T}. \tag{18}$$

Here contour $C_I$ is transformed and compared with $C_M$ in the global coordinate frame. The above objective function is an approximation to our nonlinear optimization (4) because the covariance matrix used to weight the first term in Eq. (17), which is the uncertainty of the IVUS contour in the global coordinates, will be different from $\Sigma_I$, and it cannot be directly computed since it will use the catheter pose $\mathbf{P}$, which is the state vector.

The difference between the contours in Eq. (17) can also be presented as the distance vector $\sum_{j=1}^{n} \| d_j \|^2 = \sum_{j=1}^{n} \| c_j^M - g_j(\mathbf{P}) \|^2$, and the corresponding Jacobian $J_j'$ can then be computed as

$$J_j' = -\frac{\partial d_j}{\partial g_j} \frac{\partial g_j}{\partial \mathbf{P}}. \tag{19}$$

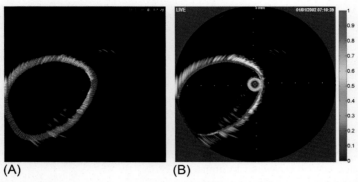

**FIG. 3** The uncertainty estimation of the IVUS contour. (A) The intensities of the points around the IVUS contour (*highlighted in red*) are accumulated. (B) The color map of the uncertainty of the IVUS contour.

The distances $\{d_j\}$ can be computed as the shortest distances from contour $g(\mathbf{P}) = [g_1(\mathbf{P})^T, ..., g_n(\mathbf{P})^T]^T$ to the 3D mesh of the vessel in the preoperative data. This can be precomputed as the distance space $\mathcal{D}$ of the preoperative data,[21] and its partial derivative can be computed as the gradient $\nabla\mathcal{D}$ of the distance space:

$$d_j = \mathcal{D}(g_j(\mathbf{P})), \quad \frac{\partial d_j}{\partial g_j} = \nabla\mathcal{D}(g_j(\mathbf{P})). \qquad (20)$$

Here a signed distance space $\mathcal{D}$ is used instead of an unsigned one since the gradient will be wrong if the point is around the mesh surface.

The optimization function formulated above is computationally efficient and enables real-time vessel reconstruction because the distance space $\mathcal{D}$ and the gradient $\nabla\mathcal{D}$ can be precalculated on the preoperative data. This means there is no need to compute the Jacobian and the contour differences in the first term of the objective function during the optimization, allowing the reconstruction to be performed in real time. The process of solving the problem by using the proposed real-time implementation is shown in the algorithm chart in Fig. 4.

To demonstrate the validity of the real-time implementation, here we give a lemma:

**Lemma 1.** *The approximated problem in Eq. (17) is equivalent to the original problem in Eq. (4), if the covariance matrix of the IVUS contour $\Sigma_I$ is spherical.*

*Proof.* The weighted 2-norm of the first term in Eq. (17) can be rewritten as

$$\| c_j^M - g_j(\mathbf{P}) \|_{\Sigma_j^{-1}}^2 = [c_j^I - f_j(\mathbf{P})]^T R \Sigma_I^{-1} R^T [c_j^I - f_j(\mathbf{P})]. \qquad (21)$$

If $\Sigma_I$ is spherical, then $R\Sigma_j^{-1}R^T = RR^T \Sigma_j^{-1} = \Sigma_j^{-1}$ and Eq. (17) is equivalent to Eq. (4).

## 4. RESULTS

Two aortic phantoms created by Materialise (Leuven, Belgium) were used for validation, one made of plexiglass (Plexiglass) and the other made of HeartPrint material (HeartPrint) (see Fig. 5). The two phantoms were first scanned by CT, and the meshes of the inner walls were segmented from the CT scan to provide the triangular surface meshes of the models (see Fig. 8). The distance space as well as the gradient of the distance space were precalculated from the 3D meshes. An Aurora 6 DoF EM sensor (NDI, Waterloo, Canada) was attached to the tip of a Visions PV 8.2 IVUS catheter (Volcano, San Diego, CA) to provide its position and orientation.

The calibration between IVUS and EM sensors on the catheter was estimated by using a batch optimization of the proposed algorithm for a prelogged dataset, by considering all the catheter poses as well as the relative pose between IVUS and EM sensors in the state vector. Since the IVUS-EM relative pose is fully correlated to all the catheter poses in the normal equations matrix, accurate results can be achieved. Markers attached to the phantoms were used to perform a rigid registration between the CT and EM coordinate systems. Alternatively, the CT-EM registration can also be done similar to the IVUS-EM calibration using the batch optimization scheme of the proposed optimization framework for a prelogged dataset. In our experiments, the CT-EM registration results provided by the two methods are similar.

### 4.1. Simulation and Robustness Assessment

Simulations were first performed to assess the influence of noise and robustness of the proposed algorithm to

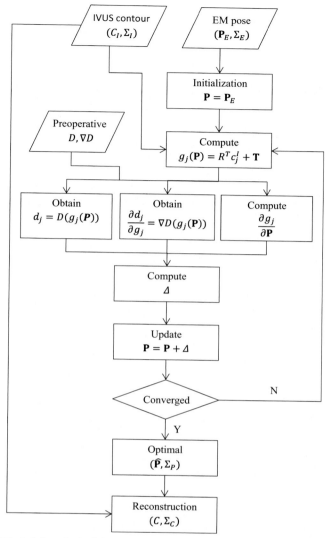

FIG. 4 A flowchart of the proposed real-time implementation of SCEM+.

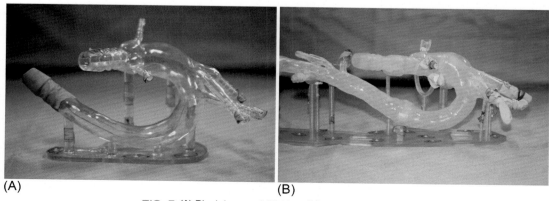

(A)                                                (B)

FIG. 5 (A) Plexiglass and (B) HeartPrint aortic phantoms.

different levels of noise in the data. Simulated vessel contours were generated from one set of preoperative data from the Plexiglass phantom along with known EM poses. Different levels of zero mean Gaussian noise were added to the EM poses and vessel contours and the results are shown in Fig. 6.

In this simulation, zero mean Gaussian noise with $\sigma_R$ = {0.02, 0.05, 0.1, 0.15, 0.2, 0.25, 0.3} rad on the rotation, $\sigma_T$ = {0.2, 0.5, 1, 1.5, 2, 2.5, 3} mm on the translation of EM poses, as well as $\sigma_C$ = {0.2, 0.4, 0.6, 0.8, 1.0, 1.2, 1.5} mm on the IVUS contours was added separately, and 10 independent runs of each noise level were generated to get a total of 210 datasets as the input of the algorithm. For each level of noise, the mean errors of the 10 runs for rotation, translation, and reconstruction of the results were computed and shown in Fig. 6 as red, green, and black lines, respectively. Here, the maximum noise levels were within the range of the errors in the real EM and IVUS systems. The maximum noise level of $\sigma_R$ and $\sigma_T$ on the EM poses was defined as 0.3 rad and 3 mm, as the accuracy of the 6DoF sensor reported for the Aurora EM System was 0.8 mm (1.2 mm 95% CI)

for position and 0.012 rad (0.014 rad 95% CI) for orientation from the user guide. The maximum noise level of $\sigma_C$ was defined as 1.5 mm since the width of the high-intensity vessel wall reflections in the IVUS images was around 1.5 mm.

From Fig. 6 we can see that, when adding noise on the rotations of EM poses, as noise increases, the error of pose translations increased accordingly while the error of the 3D reconstruction changed a little. When adding noise on the translations of EM poses, the error of 3D reconstruction was more sensitive than that of rotations. Again, when adding noise on the IVUS contour, as noise increases, the error of pose translations increased accordingly, while the error of rotations changed little. It can be seen that the errors were still very small even when large noise was added on the observations and the reconstruction accuracy was most sensitive to noise in the translations of EM poses as well as the IVUS contours.

## 4.2. Phantom Experiments

For the phantom experiments, a total of 25 datasets with different EM-Phantom setups (see Fig. 7) and 2

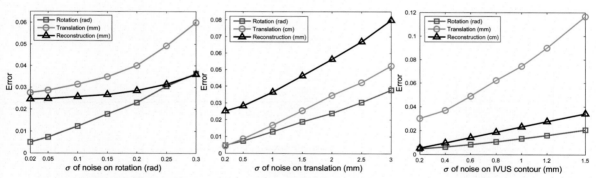

FIG. 6 Accuracy of rotation and translation of poses and 3D reconstruction w.r.t noise on rotation and translation of EM poses and extracted vessel contours.

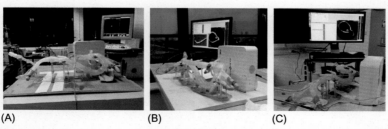

FIG. 7 Different setups at (A) KU Leuven, (B) King's College London, and (C) Northwick Park Hospital (Imperial College London).

IVUS-EM catheters were collected consisting of catheter pullbacks and/or insertions within the two phantoms. The details of the experiments are presented in Table 1. For each IVUS image frame, contour extraction was performed with a radial scan similar to that provided in Ref. 17 to identify high-intensity ultrasound reflections. The performance of the proposed reconstruction on the two phantoms was compared to SCEM[17] illustrated in Fig. 8. The significant improvement over the use of SCEM in the presence of sudden catheter motion is highlighted in Fig. 9. The trajectories of the catheter computed by the two methods are shown in Fig. 10.

Quantitative performance evaluation results are presented in Fig. 11A. The reconstruction error is calculated as the distance between the recovered structure and the preoperative data. The mean error for the HeartPrint phantom is around 0.3 mm, while that for the Plexiglass phantom is around 0.6 mm, which are significantly lower than the mean error of SCEM. The higher accuracy of the HeartPrint phantom is due to the better quality of IVUS images compared to the Plexiglass phantom.

The computational cost of the optimization algorithm presented in Fig. 11B verifies that the method can run in real time. The algorithm converged in only two to four iterations with around 0.3–0.6 ms per iteration and the optimization algorithm initialized by the pose from EM never diverged for any frame in the datasets.

To further validate the proposed vessel reconstruction, the optimized catheter poses from the proposed algorithm were used to obtain cross sections to the preoperative data, which were back projected to the original IVUS images as shown in Fig. 12. It is clear that because of the consideration of the uncertainty, even with incomplete IVUS contour extraction, an accurate pose is still estimated as the corresponding vessel contour from the preoperative model is along the high-intensity ultrasound reflections.

## 5. CONCLUSION

In conclusion, using IVUS, EM, and preoperative data, this chapter presents a real-time robust 3D vessel reconstruction and catheter navigation approach by advanced real-time nonlinear optimization technique. Compared to other methods such as SCEM, the uncertainty of IVUS and EM were used as weights in the proposed optimization algorithm which ensures that the proposed 3D vessel reconstruction was robust to contour extraction error from IVUS and sudden motion of the catheter. It was also demonstrated that a real-time implementation can be achieved with carefully selecting the coordinate frame to reformulate the objective of function of the problem. Simulation and experiments of 25 datasets with two different phantoms validated the accuracy and efficiency of the proposed algorithm.

The method simultaneously obtains the vessel structure, optimal catheter pose, and the 3D reconstruction uncertainty, which can be used for endovascular navigation. This study is validated by phantom study. If in vivo data is used, the deformation caused by the cardiac motion can be overcome by gating IVUS images with ECG data. Reconstruction of all the deformations of the vessel will be our future work. In this scenario, finite element modeling[22–25] or differential surface representations[26–28] can be combined with the current optimization framework to provide constrains of the deformation. From the vasculature, additional information can also be included a priori in the current objective function.[29–31]

### TABLE 1
### Details of the Experiments

| Datasets | Date | Place[a] | Phantom | Cath/setup[b] | Motion[c] |
|---|---|---|---|---|---|
| 1–6 | 06/2014 | Hospital | Plexiglass | C-1/S-1 | Pul |
| 7–11 | 12/2014 | Hospital | HeartPrint | C-1/S-2 | Ins/Pul |
| 12–16 | 01/2015 | Hospital | HeartPrint | C-1/S-3 | Ins&Pul+Abr |
| 17–23 | 01/2015 | KUL | HeartPrint | C-1/S-4 | Ins&Pul |
| 24–25 | 01/2015 | KUL | HeartPrint | C-2/S-4 | Ins&Pul |

[a]Hospital: Northwick Park Hospital, Imperial College London, KUL: University of Leuven.
[b]C-1: Catheter 1, S-1: Setup-1.
[c]Catheter motion: Pul(pullback), Ins(insertion) and Abr(abrupt catheter motion).

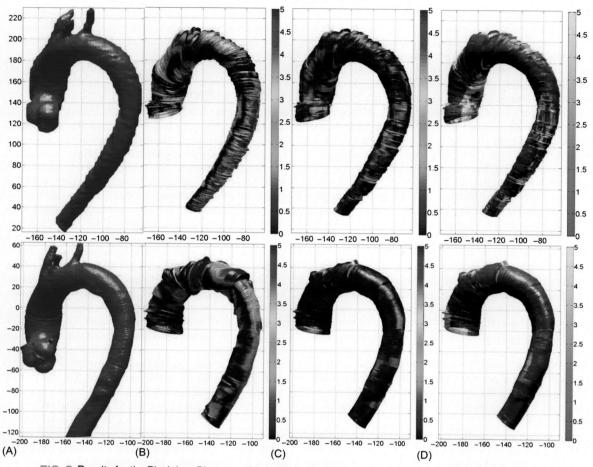

FIG. 8 Results for the Plexiglass Phantom with pullback (*first row*) and HeartPrint Phantom with insertion and pullback (*second row*): (A) the 3D mesh of the preoperative data, (B) the 3D reconstruction results by SCEM, and (C) the proposed SCEM+ algorithm with errors in mm, and (D) the uncertainty (log) map.

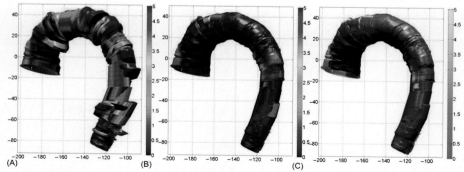

FIG. 9 Results for the HeartPrint phantom with abrupt catheter motion. (A) SCEM, (B) SCEM+, (C) uncertainty.

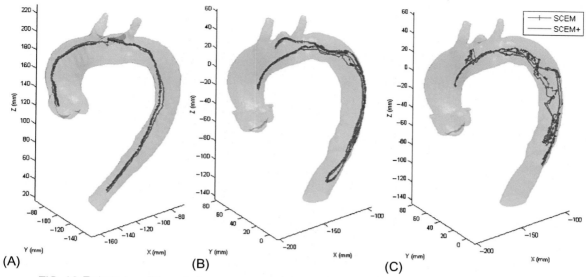

(A)                          (B)                          (C)

FIG. 10 Trajectories of the catheter computed by SCEM (*blue*) and SCEM+ (*red*). (A) Plexiglass phantom, (B) HeartPrint phantom, (C) Abrupt catheter motion.

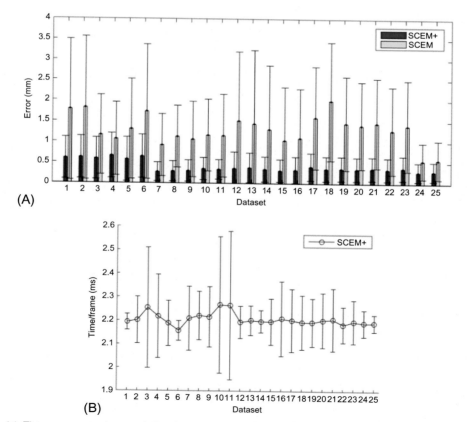

FIG. 11 The accuracy and computational cost of the proposed SCEM+ algorithm. The Plexiglass phantom was used in datasets 1–6, while the HeartPrint phantom was used in 7–25. (A) Vessel reconstruction errors; (B) computational cost.

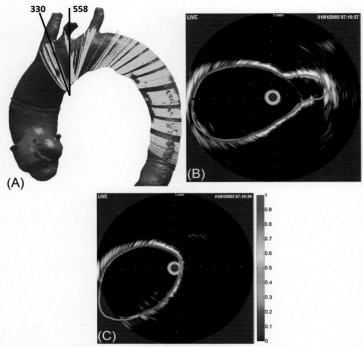

**FIG. 12** Vessel contours back projected from the preoperative mesh to IVUS images using the optimized poses from the proposed SCEM+ algorithm. The *crosses* represent the contour extracted from the IVUS images, colored by the weight in Eq. (15) (between 0 and 1). The *green line* represents the back-projected contours. (A) 3D contours, (B) IVUS image 330, (C) IVUS image 558.

## REFERENCES

1. May J, White GH, Waugh R, et al. Life-table analysis of primary and assisted success following endoluminal repair of abdominal aortic aneurysms: the role of supplementary endovascular intervention in improving outcome. *Eur J Vasc Endovasc Surg.* 2000;19(6):648–655.
2. Avino AJ, Bandyk DF, Gonsalves AJ, et al. Surgical and endovascular intervention for infrainguinal vein graft stenosis. *J Vasc Surg.* 1999;29(1):60–71.
3. Funaki B. Endovascular intervention for the treatment of acute arterial gastrointestinal hemorrhage. *Gastroenterol Clin North Am.* 2002;31(3):701–713.
4. Babaliaros V, Block P. State of the art percutaneous intervention for the treatment of valvular heart disease: a review of the current technologies and ongoing research in the field of percutaneous valve replacement and repair. *Cardiology.* 2007;107(2):87–96.
5. Vahanian A, Alfieri OR, Al-Attar N, et al. Transcatheter valve implantation for patients with aortic stenosis: a position statement from the European Association of Cardio-Thoracic Surgery (EACTS) and the European Society of Cardiology (ESC), in collaboration with the European Association of Percutaneous Cardiovascular Interventions (EAPCI). *Eur Heart J.* 2008;29(11):1–8.
6. Walther T, Chu MW, Mohr FW. Transcatheter aortic valve implantation: time to expand? *Curr Opin Cardiol.* 2008;23(2):111–116.
7. Walther T, Falk V, Kempfert J, Borger MA, Fassl J, Chu MW. Transapical minimally invasive aortic valve implantation; the initial 50 patients. *Eur J Card Thorac Surg.* 2008;33(6): 983–988.
8. Zahn R, Schiele R, Kilkowski C, et al. Correction of aortic regurgitation after transcatheter aortic valve implantation of the medtronic corevalvetm prosthesis due to a too-low implantation, using transcatheter repositioning. *J Heart Valve Dis.* 2011;20(1):64–69.
9. Davidson MJ, White JK, Baim DS. Percutaneous therapies for valvular heart disease. *Cardiovasc Pathol.* 2006;15:123–129.
10. Burgner J, Herrell SD, Webster R. Toward fluoroscopic shape reconstruction for control of steerable medical devices. In: *Proc ASME 2011 Dynamic Systems and Control Conference;* 2011:791–794.

11. Rosenfield K, Losordo DW, Ramaswamy K, et al. Three-dimensional reconstruction of human coronary and peripheral arteries from images recorded during two-dimensional intravascular ultrasound examination. *Circulation.* 1991;84(5):1938–1956.

12. Sanz-Requena R, Moratal D, García-Sánchez DR, Bodí V, Rieta JJ, Sanchis JM. Automatic segmentation and 3D reconstruction of intravascular ultrasound images for a fast preliminary evaluation of vessel pathologies. *Comput Med Imaging Graph.* 2007;31(2):71–80.

13. Wahle A, Prause GPM, DeJong SC, Sonka M. Geometrically correct 3-D reconstruction of intravascular ultrasound images by fusion with bi-plane angiography-methods and validation. *IEEE Trans Med Imaging.* 1999;18(8): 686–699.

14. Godbout B, Guise JAD, Soulez G, Cloutier G. 3D elastic registration of vessel lumen from IVUS data on biplane angiography. In: *Proc International Conference on Medical Image Computing and Computer Assisted Intervention (MICCAI);* 2003:303–310.

15. Bourantas CV, Papafaklis MI, Athanasiou L, et al. A new methodology for accurate 3-dimensional coronary artery reconstruction using routine intravascular ultrasound and angiographic data: implications for widespread assessment of endothelial shear stress in humans. *EuroIntervention.* 2013;9(5):582–593.

16. Bourantas CV, Kourtis IC, Plissiti ME, et al. A method for 3D reconstruction of coronary arteries using biplane angiography and intravascular ultrasound images. *Comput Med Imaging Graph.* 2005;29:597–606.

17. Shi C, Giannarou S, Lee SL, Yang GZ. Simultaneous catheter and environment modeling for trans-catheter aortic valve implantation. In: *Proc IEEE/RSJ International Conference on Intelligent Robots and Systems (IROS);* 2014:2024–2029.

18. Shi C, Tercero C, Ikeda S, et al. In vitro three-dimensional aortic vasculature modeling based on sensor fusion between intravascular ultrasound and magnetic tracker. *Int J Med Rob Comput Assisted Surg.* 2012;8(3):291–299.

19. Hartley R, Zisserman A. *Multiple View Geometry in Computer Vision.* Cambridge, UK: Cambridge University Press; 2003.

20. Morton KW, Mayers DF. *Numerical Solution of Partial Differential Equations: An Introduction.* Cambridge, UK: Cambridge University Press; 2005.

21. Jones M, Baerentzen JA, Sramek M. 3D distance fields: a survey of techniques and applications. *IEEE Trans Vis Comput Graph.* 2006;12(4):581–599.

22. Celniker G, Gossard D. Deformable curve and surface finite-elements for free-form shape design. In: *Proc ACM SIGGRAPH;* 1991:257–266.

23. Cirak F, Ortiz M, Schroder P. Subdivision surfaces: a new paradigm for thin-shell finite-element analysis. *Int J Numer Methods Eng.* 2000;47(12):2039–2072.

24. Cirak F, Scott M, Schroder P, Ortiz M, Antonsson E. Integrated modeling, finite-element analysis, and design for thin-shell structures using subdivision. *Comput Aided Des.* 2002;34(2):137–148.

25. Thomaszewski B, Wacker M, Strasser W. A consistent bending model for cloth simulation with corotational subdivision finite elements. In: *Proc ACM SIGGRAPH/ Eurographics Symposium on Computer Animation;* 2006: 107–116.

26. Botsch M, Sorkine O. On linear variational surface deformation methods. *IEEE Trans Vis Comput Graph.* 2008;14(1):213–230.

27. Botsch M, Sumner R, Pauly M, Gross M. Deformation transfer for detail-preserving surface editing. In: *Proc Vision, Modeling, and Visualization (VMV);* 2006:357–364.

28. Yu Y, Zhou K, Xu D, et al. Mesh editing with Poisson-based gradient field manipulation. *ACM Trans Graph.* 2004;23 (3):644–651.

29. Torresani L, Hertzmann A, Bregler C. Nonrigid structure-from-motion: estimating shape and motion with hierarchical priors. *IEEE Trans Pattern Anal Mach Intell.* 2008;30(5):878–892.

30. Dai Y, Li H, He M. A simple prior-free method for non-rigid structure-from-motion factorization. *Int J Comput Vis.* 2014;107(2):101–122.

31. Terzopoulos D, Witkin A, Kass M. Constraints on deformable models: recovering 3D shape and nonrigid motion. *Artif Intell.* 1988;36(1):91–123.

# Index

Note: Page numbers followed by *f* indicate figures and *t* indicate tables.

Printed in the United States
By Bookmasters